T0295367

Manual for Bryophytes

Manual for Bryophytes
Morphotaxonomy, Diversity, Spore Germination, Conservation

Samit Ray
Shewli Bhattacharya

CRC Press
Taylor & Francis Group
Boca Raton London New York

CRC Press is an imprint of the
Taylor & Francis Group, an **informa** business

LEVANT
Levant Books
Kolkata

First published 2022
by CRC Press
2 Park Square, Milton Park, Abingdon, Oxon, OX14 4RN

and by CRC Press
6000 Broken Sound Parkway NW, Suite 300, Boca Raton, FL 33487-2742

© 2022 Samit Ray, Shewli Bhattacharya and Levant Books

CRC Press is an imprint of Informa UK Limited

The right of Samit Ray and Shewli Bhattacharya to be identified as authors of this work has been asserted by them in accordance with sections 77 and 78 of the Copyright, Designs and Patents Act 1988.

All rights reserved. No part of this book may be reprinted or reproduced or utilised in any form or by any electronic, mechanical, or other means, now known or hereafter invented, including photocopying and recording, or in any information storage or retrieval system, without permission in writing from the publishers.

For permission to photocopy or use material electronically from this work, access www.copyright.com or contact the Copyright Clearance Center, Inc. (CCC), 222 Rosewood Drive, Danvers, MA 01923, 978-750-8400. For works that are not available on CCC please contact mpkbookspermissions@tandf.co.uk

Trademark notice: Product or corporate names may be trademarks or registered trademarks, and are used only for identification and explanation without intent to infringe.

Print edition not for sale in South Asia (India, Sri Lanka, Nepal, Bangladesh, Pakistan or Bhutan).

British Library Cataloguing-in-Publication Data
A catalogue record for this book is available from the British Library

Library of Congress Cataloging-in-Publication Data
A catalog record has been requested

ISBN: 978-1-032-07694-2 (hbk)
ISBN: 978-1-003-20837-2 (ebk)

Typeset in Minion Pro 10.5
by Levant Books, Kolkata 700014

LEVANT

The book is dedicated to
Late Dr. Probir Chatterjee
The teacher who gave me early lessons on bryophytes

Preface

This book is dedicated to Late Dr. Probir Chatterjee who was a teacher and researcher per excellence, a naturalist and a great motivator for the students. Though he was known in India as an algologist, Dr. Chatterjee had devoted much of his time in studying bryophyte diversity, collecting specimens and teaching bryophytes in under-graduate and post-graduate classes. It is unfortunate that, in recent times, the subject of Bryology has not got its due importance in the teaching and research curricula in most of the Universities. In spite of the great legacy of Late Prof. H. C. Gangulee, who wrote a series of monographs on mosses from Eastern Himalaya and its adjacent areas, the study of bryophytes has not received much attention in this part of India. It is only in a few centres and universities that research work is being carried out on bryophytes. National Botanical Research Institute, Lucknow and Botanical Survey of India have been contributing significantly in bryophyte research.

There is a great significance of bryophytes in land plant evolution, evolution of gene and gene function, molecular phylogeny, desiccation tolerance, geochemical cycle, land colonization and medicinal use. Also, the role of bryophytes with regard to water retention, prevention of soil erosion, nutrient cycling, nitrogen fixation and pollution monitoring is immense. Considering the above and the occurrence of a very rich bryophyte flora in India, I found it necessary to draw the attention of students and researchers to this unique group of plants. Following the footsteps of Late Dr. Chatterjee, I have been teaching Bryology since the very beginning of my teaching career in 1986. I also made it a point to collect sample specimens from various locations, particularly from Eastern Himalaya and from certain parts of Eastern India, to demonstrate to the students, morphological and structural variations of bryophytes and preserve voucher specimens.

After the untimely death of Dr. Probir Chatterjee, the students, colleagues, admirers of Dr. Chatterjee formed "Probir Chatterjee Research Foundation" to augment teaching and research in Cryptogamic Botany. The foundation organizes workshops where various practical aspects are demonstrated to the students, researchers and young teachers. While conducting such workshops it was felt necessary that young students should have a manual as a practical guide to study bryophytes. Earlier, in 2013, the foundation had published on its own initiative a manual covering several aspects of Cryptogamic Botany. The present manual that concentrates only on bryophytes will provide guidelines for collection and recording of bryophytes, methodologies for studying the morphology and internal structure of bryophytes, understanding taxonomic significance of morphological features and the need for conserving the bryoflora. Hope this manual will draw the attention of the students and young researchers to the world of bryophytes and help them in their pursuit of studying the natural vegetation of bryophyte.

Samit Ray

Foreword

It is well established that bryophytes contribute to a great extent in terrestrial ecology, pollution monitoring and have played a significant role in land plant evolution. In spite of such crucial role in nature this particular group of plants has not received due attention like their sister embryophytes on land – the trachaeophytes, probably due to their usually *incognito* presence in most of the ecoregions. However, with increasing awareness, their role in natural environment is being appreciated and there has been a great advancement in various fields of bryological researches, like biosystematics, molecular phylogeny, biochemical pathways, mechanism of desiccation tolerance, ecophysiology, population dynamics and the relationship between bryophyte and the climate change. These studies have helped us understand the basic biology of bryophytes up to the molecular level. But for a beginner in the field of bryology, understanding of the basic methods of collection, preservation and identification of common bryophytes and their key morphological characteristics are of prime importance. As bryophytes show great phenotypic plasticity depending on the variations of ecological conditions to which they are exposed to, a clear idea of morphology and its correlation with ecological parameters is necessary for their easy identification.

It is pertinent to mention here that beginning with the 'father of bryology in India' – late Prof. Shiv Ram Kashyap in early twentieth century, there were many eminent bryologists in India, like Prof. S.K. Pande, Prof. P.N. Mehra, Prof. R.S. Chopra, Dr. D.C. Bharadwaj, Prof. Ram Udar, Prof. P. Kachroo, Prof. S.S. Kumar, Prof. R.N. Chopra, Prof. S.C. Srivastava, Prof. Dinesh Kumar and others, who have immensely contributed to our knowledge of various facets of Indian bryology, like morpho-taxonomy, floristics, phytogeography, developmental morphology, cytology, phylogeny, etc. For long time the bryological researches in India remained confined to the centers at Punjab University, Lahore/Chandigarh, Lucknow University, National Botanical Research Institute, Lucknow, Delhi University, Calcutta University and the Botanical Survey of India, thus limiting the geographical area coverage of these studies. While, the last few decades have witnessed the sprouting up of several new centers of bryological researches in different parts of the country and increasing interest among the young botanists in these plants, gradual decline of taxonomy from the curricula of under-graduate and post-graduate courses of Botany across the Indian universities has left the students, opting for a carrier in the field of plant science research including bryophytes, grappling with the basic issues like identification of plant species.

It is in this context, the '*Manual for Bryophytes: Morphotaxonomy, diversity, spore germination, conservation*' by Prof. Samit Ray and Dr. Shewli Bhattacharya to introduce this fascinating group of plants to students and young teachers is really commendable. In the introductory chapter the authors have enumerated the fundamental features of bryophytes which will help students understand the position of the bryophytes in the plant kingdom. A comprehensive account of classification of bryophytes in Chapter II and IV will help students to understand the morphological variability in bryophytes and its significance in classification of these plants into Class, sub-class, order and families. Methodologies related to collection, preservation, study of morphology, anatomy, preparation of slide, spore morphology and spore germination are well described. Specifically, the study of spore morphology through SEM and other dimensional aspects are well appreciated. Line drawings and related photomicrographs of commonly occurring genera of liverworts, hornworts and mosses will be of great advantage to the students and young teachers. The final chapter on bryophyte conservation will help create awareness about the effect of environmental degradation on the natural populations of bryophytes and its amelioration.

The book is written in lucid language and the text is properly divided into subheadings to make it student friendly. The book will be of immense use to students and young teachers as a practical guide. I congratulate the authors for bringing out this wonderful companion of students of bryology.

Dr. D.K. Singh
Scientist G (Retd.)
Botanical Survey of India
Member, Advisory Committee
on Ecosystem Research Programme (EcR
Ministry of Environment, Forests & Clim
Change Govt. of India

Acknowledgements

We gratefully acknowledge the assistance of Dr. D. K. Singh, Former Scientist 'G' of Botanical Survey of India and one of the finest workers on bryoflora of India. He has written the foreword for this manual, given valuable suggestions after going through the draft manuscript and provided several photographs of liverworts, hornworts and mosses. Dr. Md. N. Aziz, Scientist 'E' Cryptogamic Unit and Dr. D. Singh, Scientist 'C' CNH, Botanical Survey of India, has assisted us in identifying some mosses and liverworts. Prof. S. P. Adhikary, President, Prof. Ruma Pal, Joint-Secretary and other members of PCRF have constantly encouraged and motivated us in preparing this manuscript. We also thank our colleagues and friends for their suggestions and encouragement. Prof. Samit Ray is indebted to his wife Smt. Nandini Ray for her willing cooperation and encouragement. Dr. Shewli Bhattacharya is indebted to her parents and husband for their support and encouragement. Lastly, we take this opportunity to thank our publisher Levant Books, Kolkata for agreeing to publish this manual.

Samit Ray

Contents

List of Plates

Chapter 1

Introduction

The bryophytes, commonly referred to as the amphibians of the plant kingdom, are the most diverse of the land plants after Magnoliophyta (3,50,000 species). There are approximately 19,000 species (6000 – 8000 species of liverworts, 100 - 200 species of hornworts and 10,000 – 15,000 species of mosses). However, opinions vary on the number of species. According to Gradstein et al. (2001) there are 15,000 species while Crum (2001) put the number as 25,000. These organisms share a fundamentally similar life-cycle with a perennial and free-living, photosynthetic gametophyte alternating with a short-lived sporophyte that completes its entire development attached to the maternal gametophyte. All the three groups comprise the earliest lineages of land plants derived from green algal ancestors (the charophycean green algae belonging to Streptophyta). It is well established from the structural and morphological variability and molecular phylogenetic study that the bryophytes do not constitute a single monophyletic lineage and divided into three divisions – Marchantiophyta, Anthocerotophyta and Bryophyta. The bryophytes, according to morphological cladistics analyses are not monophyletic (Mishler and Churchill 1984, 1985; Mishler et al. 1994; Kenrick and Crane, 1997). Recent, molecular cladistics analyses also confirm non-monophyly. The most accepted view of interrelationship between three major bryophyte lineages is that of Qiu et al. (2006). Combining an extensive set of DNA sequences of representatives of the major lineages, it was proposed that hornworts share a common ancestor with vascular plants, whereas liverworts are a sister lineage to all other extant embryophytes. Mosses bridge the gap between liverworts and hornworts.

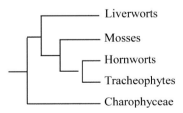

Fig. 1.1. Phylogenetic relationship based on gene sequence data and genomic structural characters (Kelch et al. 2004, Qiu et al. 2006).

Bryophytes are characterized by the following features – (i) they are referred to as amphibians of the plant kingdom because of their dependence on water for luxuriant growth and also sexual reproduction, (ii) bryophytes are unique among land plants in that they possess an alternation of generations which involves a dominant free-living, haploid gametophyte alternating with reduced dependent diploid sporophyte, (iii) the gametophyte may be thallose or foliose and exhibit wide range of developmental, structural and morphological variations, (iv) lignified vascular tissue, meristematic tissue and secondary growth are absent, (v) most of the bryophytes are poikilohydric (poikilohydrous organisms maintain equilibrium with atmospheric humidity, gaining or loosing water readily), (vi) some bryophytes do contain vascular tissues including highly specialized water conducting cells. These cells do not form lignified walls. Among the liverworts an internal strand of specialized water conducting cells occurs in gametophytes of Calobryales and in some members of Metzgeriales. These are dead cells without cytoplasmic content at maturity. In the sub-class Bryidae water conducting system consists of elongate cells, lacking cytoplasmic content at maturity called hydroids. The hydroids associated with stereides (thick walled living cells) form a central strand in the leafy stem of the gametophyte sometimes referred to as hydrom. Hydroids do not contain lignin and lack secondary wall pattern. In the members of polytrichales, besides hydroids, specialized cells with marked morphological similarity to the protophloem sieve cells in trachaeophytes are found called leptoids. They occur both in gametophyte and sporophytic seta. Leptoids have been found to be preferential route for the translocation of organic nutrients, (vii) the male sex organ antheridium producing biflagellate antherozoids and female sex organ archegonium containing egg are multicellular structures where fertile cells are surrounded by sterile jacket, (viii) delayed meiosis of zygote and interpolating mitotic cell divisions result in the production of an embryo, (ix) the embryo differentiates within the venter of the archegonium and ultimately develops into a sporophyte, (x) there are distinct embryonic regions determined to develop into the three organotrophic zone of the mature sporophyte – foot, seta and capsule, (xi) within the capsule of sporophyte meiotic cell divisions take place to produce spores; spores are homosporous, (xii) this sporophyte is monosporangiate and matrotrophic in growth and parasitic on gametophyte, (xiii) spores germinate to form a protonemal stage; the protonemal stage in the liverworts is globose while in the mosses the protonema is filamentous and more conspicuous in the life-cycle.

In the Devonian period, a diversity of plants adapted to the terrestrial environment and were able to absorb water, nutrients. Absorbed water and nutrients were transported and distributed throughout their aerial shoots. The ability to exist on land is the result of numerous complex interactions that involved several structural and physiological adaptations – (i) anchorage and water uptake, (ii) structural support and water transport, (iii) protection against desiccation and radiation, (iv) gaseous exchange, (v) reproduction on land.

On the basis of a variety of ultrastructural, biochemical and molecular data Duckett and Renzaglia (1988) suggested that the principal bryophyte groups had separate origins and that hornworts, liverworts and mosses represent the earliest divergent lineages of extant land plants. Phylogenetic evidence suggests that bryophytes in general and liverwort like plants in particular should have been important components of early terrestrial floras (Bateman et al. 1998; Renzaglia et al. 2007).

The earliest recognizable bryophytes in the macrofossil record include liverworts that appear to have their closest affinities with the Metzgeriales. The earliest liverwort in the fossil record is *Metzgeriothallus sharonae* from Givetian (upper middle Devonian) shales and siltstones from New York (Van Aller Hernick et al. 2008). There are some records of mosses as early as Carboniferous (Walton 1928). One of these is *Muscites plumatus,* an impression of a small leafy shoot in rocks of Mississipian age (Thomas, 1972). The earliest macrofossil that bears some resemblance to a modern hornwort is *Dendroceros victoriensis* from the lower Cretaceous Koonwarra fossil bed in Australia (Drinnan and Chambers, 1986).

Bryophytes inhabit a very wide range of ecosystems, habitats, microhabitats including substrates on which vascular plants cannot live. It must be noted that in most of the environments vascular plants are the dominant vegetation. However, in arctic, the antarctic, in alpine habitats, in mountains above tree line, in bogs, in fens and larger peatlands bryophytes are often the dominant plants in terms of both biomass and productivity. It is not only that they prefer some specific ecological conditions conducive for their growth but they also have important ecological functions beneficial for the living world. Some of their ecological significance are briefly mentioned here – (i) extensive carpets of bryophytes on soil help moisture conservation, absorb and retain water and this absorbed water can be released for a long period, (ii) they are good in trapping nutrients from air and also absorb nutrients carried by water. These nutrients are available after the bryophytes die and decay, (iii) nutrition is also supplemented by N_2-fixation where *Nostoc* colonies grow in symbiotic association with genera like *Anthoceros, Blasia,* (iv) some species are very good soil binders and maintain soil structure under the crust, (v) bryophyte colonies provide niches for numerous invertebrates which form a part of food chain for higher organisms, (vi) bryophytes are often the first plants to colonize barren surfaces like road cuttings, road out ropes and volcanic ash, (vii) moss carpets also help in seed germination for higher plants by providing a moist seed bed.

The world wide reduction, fragmentation and degradation of habitats important for bryophytes have led to loss of species richness and genetic diversity. Fortunately conservation of bryophytes has been emphasized in the United Nations Convention of Biological Diversity. One of the significant aspects of the action plan is taking initiative to create inventories to determine bryophyte richness in different regions and habitat types and to determine which species are locally common, rare or threatened. Another important aspect is comparing bryophyte floras from undisturbed and disturbed habitats to determine the impact of disturbance and to identify those species unable to survive in disturbed areas. Both these aspects need good knowledge of morphology of bryophytes.

Bryophytes show extensive phenotypic plasticity. Depending on the ecological parameters, taxonomically important morphological and structural features change. So there must be a thorough study of the correlation between ecological parameters of the habitat and the morphological features of a particular bryophyte species. There is another aspect of this kind of study. Bryophyte species tend to be highly specific for particular microenvironment and respond to factors as temperature, light, water availability, substrate chemistry etc. Thus these plants can be good ecological indicator species.

Chapter 2

Classification of Bryophytes

The names Bryophyta and Pteridophyta were introduced by Haeckel (1866) but he never gave them rank of a division. Schimper (1879) treated Bryophyta as a division of the plant kingdom. The system of classification in which cryptogamic portion of the plant kingdom is placed in three divisions – Thallophyta, Bryophyta and Pteridophyta first appeared in Eichler's "Syllabus...." (1883). Division Thallophyta included the classes Algae and Fungi and Division Bryophyta was divided into classes – Hepaticae and Musci. This system of classification helped the botanists of that era to distinguish between algae and bryophytes. Engler (1892) recognized Hepaticae and Musci as two classes and divided Hepaticae into three orders – Marchantiales, Jungermanniales and Anthocerotales and Musci also into three orders – Sphagnales, Andreaeales and Bryales. After detail study, Anthocerotales, usually listed as an order of Hepaticae, was given an isolated position by Underwood (1894). Howe (1899) raised the order Anthocerotales, containing *Anthoceros* and related genera, to the rank of a class and divided the division Bryophyta into three classes – Hepaticae, Anthocerotes and Musci. This system was followed by all leading workers like Smith (1938, 1955), Takhtajan (1953) and Schuster (1958). The name Anthocerotes was changed to Anthocerotae.

Cavers (1911) omitted the words Hepaticae and Musci and divided Bryophyta into ten orders – Sphaerocarpales, Marchantiales, Jungermanniales, Anthocerotales, Sphagnales, Andreaeales, Tetraphidales, Polytrichales, Buxbaumiales and Eubrayles. But this classification ignores the fundamental difference between the Hepaticae and the Musci and does not seem to be justified.

Rothmaler (1951) changed the nomenclature of the three classes of Bryophyta to Hepaticopsida, Anthoceropsida and Bryopsida, which are in accordance with the recommendations of International code of Botanical Nomenclature, Utrecht 1956. Proskauer (1957) was of opinion that the name Anthoceropsida be changed to Anthocerotopsida. The differences between the Hepaticae and Musci are so great that some taxonomists want to raise them to divisions. Bold (1956) gave the status of divisions to the classes Hepaticae and Musci and they were named Hepaticophyta and Bryophyta respectively.

The placement of liverworts, hornworts and mosses in a single division, Bryophyta, carries the message that these three groups of plants are closely related to one another phylogenetically. The bryophytes are very similar in life-cycle pattern, the plant bodies in each case a free living gametophyte on which a partially or completely parasitic sporophyte is borne (the relative balance between the two generations are similar). However, the morphological, anatomical and developmental differences among them are sufficiently significant to revise the classification of the liverworts, hornworts and mosses and elevate them as separate divisions of the plant kingdom – Hepatophyta, Anthocerotophyta and Bryophyta (Stotler and Crandall-Stotler, 1977 and Crandall-Stotler, 1980). Bold et al. (1987) adopted similar classification in their book "Morphology of Plants and Fungi".

The divergent conclusions regarding their origin and phylogenetic relationship also supported such a classification. Characteristics such as biflagellate sperms, multicellular sex-organs, sporic meiosis, aerial spore dissemination and the terrestrial habitat may seem significant and important common attributes between these three groups but all these characters also occur in the algae or in the vascular plants.

Stotler and Crandall-Stotler (1977) divided the Hepatophyta into three classes - Marchantiopsida, Jungermanniopsida and Haplomitriopsida. Division Anthocerotophyta had a single class Anthocerotopsida and the Division Bryophyta was divided into three classes – Sphagnopsida, Andreaeopsida and Bryopsida. Presently, the division Bryophyta is divided into classes – Takakiopsida, Sphagnopsida, Andreaeopsida, Andreaeobryopsida, Oediopsida, Tetraphidopsida, Polytrichopsida, Bryopsida (Goffinet, Buck and Shaw, 2009). Modern studies of cell structure and molecular biology confirm that bryophytes comprise three separate evolutionary lineages which are recognized as liverworts (Marchantiophyta – this divisional name based on type species *Marchantia polymorpha*), hornworts (Anthocerotophyta), and mosses (Bryophyta). Kenrick and Crane (1998) proposed that the three divisions (groups) of bryophytes represent a grade or structural level in plant evolution, identified by their monosporangiate life-cycle. Detail characteristic features of the three divisions are as follows :

Marchantiophyta – Dorsiventrally differentiated (sometimes radial and erect) gemetophytes are either simple thallose or with leaf-like appendages (foliose), leaves are lobed and without midvein, internal tissues are homogeneous or differentiated into photosynthetic and storage region, oil bodies are present in the cell, rhizoids are hyaline and one-celled, sex organs are borne terminally (in leafy forms) or developed from superficial dorsal cells, sporophyte is simple or differentiated into foot, seta and capsule, it is always of limited growth, seta elongates prior to spore release, sporogenous cells of endothecial origin, capsules contain spirally thickened elaters but no columella, capsule dehisces into four valves.

Anthocerotophyta – Gametophyte dorsiventrally differentiated, simple thallose, not foliose, internal tissue homogeneous, oil bodies present and cells contain single plastid with pyrenoid, rhizoids are hyaline and one-celled, stomata present in both gametophyte and sporophyte, sex organs are embedded in gametophyte – archegonium formed from superficial and antheridium from hypodermal cells, sporophyte differentiated into capsule and foot, no seta present, sporophyte is of indeterminate growth by a basal meristematic zone situated in between foot and capsule, sporogenous cells are of amphithecial origin, endothecium forms the sterile columella, pseudoelaters present, capsule dehisces into two valves.

Bryophyta – Gametophyte is represented by a leafy shoot, usually of radial symmetry, leaves (phyllids) are arranged on stem-like caulid, leaves are not lobed and with a midvein, water conducting cells are present in both gametophyte and sporophyte, rhizoids are brown and multicellular, stomata present on sporophyte capsule, sex organs develop from superficial cels at the top of apical or axillary meristems, sporophyte differentiated into foot, seta and capsule, seta is emergent early in the development of sporophyte, sporogenous cells of amphithecial or endothecial origin, columella is always present, capsule dehisces by operculum and peristome (in peristomate mosses).

Details of classification (after Crandall-Stotler, Stotler and Long, 2009; Goffinet, Buck and Shaw, 2009; Renzaglia, Villarreal and Duff, 2009)

MARCHANTIOPHYTA Stotler & Crand.-Stotl.

Class – Haplomitriopsida Stotler & Crand.-Stotl.

Plants bearing foliar appendages at nodes; stem axes secreting copious mucilage from epidermal cells; association with glomeromycotan fungi, apical cell tetrahedral, antheridia and archegonia loosely organized, early antheridial development involves one primary androgonial cell, archegonial neck with four vertical rows of cells, both acro- and anacrogynous development of archegonia, sporophyte large, enclosed by a fleshy shoot calyptra.

Subclass – Treubiidae Stotler & Crand.-Stotl.

Thallus prostrate, dorsiventrally flattened; leaves in two rows, unequally divided into a small dorsal lobule and large ventral lobe, longitudinal or slightly succubous; oil bodies large in specialized cells; rhizoids ventral, scattered; gemmae multicellular, not in receptacles; antheridia and archegonia protected by dorsal lobules; capsule ovoidal, wall 3-5 stratose, dehiscence 4-valved.

Order – Treubiales Schljakov

Subclass – Haplomitriidae Stotler & Crand.-Stotl.

Thallus differentiated into branched leafless stolon and erect leafy shoots; leaves in three rows, third row dorsal, transverse or slightly succubous, iso- or anisophyllous; stem with a central strand of thin walled cells; oil bodies small, homogeneous in all cells; rhizoids absent; antheridia and archegonia scattered on stem and leaf axils; capsule cylindrical, wall unistratose, dehiscence along 1-4 sutures, non-valvate.

Order – Calobryales Hamlin

Class–Marchantiopsida Cronquist, Takht & W. Zimm.

Members have thalloid (complex) gametophyte, a dorsal tilt of the apical cell which is cuneate with four cutting faces, four primary androgonial cells formed during antheridial development, six rows of archegonial neck cells, embryo octamerous, a unistratose capsule wall, unlobed spore mother cell.

Subclass – Blasiidae He-Nygrén, Juslén, Ahonen, Glenny & Piippo

Thallus not differentiated dorsiventrally, margin scarcely to deeply lobed, lobes longitudinal in insertion, air chambers or air pores absent; oil bodies absent or few in unspecialized cells; ventral scales in two rows without appendages, endogenous colonies of *Nostoc* filaments in ventral auricles; multicellular gemmae present in cups; antheridia in two rows, partially embedded; sporophyte dorsal at thallus apex; involucre tubular; seta elongate, massive, capsule wall 2-4 stratose, dehiscence by 4 valves.

Order – Blasiales Stotler & Crand.-Stotl.

Subclass - Marchantiidae Engl.

Thallus differentiated dorsiventrally, air chambers and air pores present or absent in few genara; specialized oil-cells with oil-bodies; ventral scales present or absent, appendages present or absent; rhizoids smooth walled or pegged or both; multicellular gemmae present in specialized gemma cups; antheridia may be embedded in the thallus on dorsal side or in cushion or elevated on stalked receptacles (antheridiophores); archegonia on stalked receptacles in most of the genera; sporophytes on stalked receptacles or borne dorsally or embedded, involucre present or rarely absent, seta short or moderate in length, elaters present in most spcies, capsule dehiscence by longitudinal valve or slit, sometimes by degeneration of wall.

Order – Sphaerocarpales Cavers, **Neohodgsoniales** D.G. Long,
Lunulariales D.G. Long**, Marchantiales** Limpr.

Class – Jungermanniopsida Stotler & Crand.-Stotl.

Thalloid (simple) or leafy forms of the gametophyte, apical cell tetrahedral in leafy forms; lenticular in one line of simple thalloid; tetrahedral, cuneate in other lineages, antheridial development involves formation of two primary androgonial cell, five vertical rows of cells in archegonial neck, embryo is filamentous, capsule wall composed of two or more layers of cells, spore mother cell lobed.

Simple thalloid liverworts are now interpreted as a paraphyletic grade composed of the Pelliidae (Simple thalloid I) and the Metzgeriidae (Simple thalloid II). The leafy liverworts are put into Jungermaniidae – split into two lineages – Porellales (Leafy I) and Jungermanniales (Leafy II).

Subclass - Pelliidae He-Nygrén, Juslén, Ahonen, Glenny & Piippo

Plants thalloid without air-chambers; if leafy, leaves developing from 1-initial, never lobed, arranged in two rows, succubous; branches exogenous; antheridia on the dorsal surface of the midrib; archegonia usually anacrogynous on the dorsal surface of the midrib or stem; some taxa with a central strand containing dead water-conducting cells within the thallus mid-region.

Orders – Pelliales He-Nygrén, Juslén, Ahonen, Glenny & Piippo
Fossombroniales Schljakov, Pallaviciniales W. Frey & M. Stech

Subclass- Metzgeriidae Barthol.-Began

Plants mostly thalloid without air-chambers; if leafy, leaves developing from 3-initials, arranged in 2-rows; apical cell lenticular; branches exogenous or endogenous in origin; antheridia on abbreviated lateral or ventral branches; archegonia acrogynous on abbreviated lateral or ventral branches; capsule dehiscence 4-valved; no internal differentiation.

Orders – Pleuroziales Schljakov, **Metzgeriales** Chalaud

Subclass- Jungermanniidae Engl.

Plants leafy, isophyllous or anisophyllous with the ventral leaves (underleaves or amphigastria) smaller and morphologically different from the lateral leaves, very rarely thalloid; leaves in 2 or 3 rows, with the third row ventral, developing from 2 primary leaf initials, divided into 2 or more lobes; antheridia in the axils of modified leaves; archegonia acrogynous, usually surrounded by perianth and modified leaves; capsules variable in shape, 2-10 stratose wall and dehiscence into 4-valves.

Order- Porellales Schljakov, **Ptilidiales** Schljakov,
Jungermanniales H. Klinggr.

ANTHOCEROTOPHYTA Rothm. ex Stotler & Crand.-Stotler

The division Anthocerotophyta contains one class Anthocerotopsida. However, Renzaglia et al. (2009) divided it into two classes Leiosporocerotopsida (Genus – *Leiosporoceros*) and Anthocerotopsida.

Class – Leiosporocerotopsida Stotler & Crand.-Stotler *emend.* Duff et al.

Order – Leiosporocerotales Hässel

Thalli solid but with schizogenous cavities in old thalli; *Nostoc* colonies in longitudinally oriented strands in mucilage filled schizogenous canals; chloroplast 1 – per cell without pyrenoid; antheridia 80 per chamber; capsule with stomata; massive sporogenous tissue; pseudoelaters usually unicellular, thick-walled; sporetetrad bilateral alterno-opposite; spore yellow.

Class – Anthocerotopsida de bary *ex* Jsncz. *corr.* Prosk.

Subclass – Anthocerotidae Rosenv. *corr.* Prosk.

Thalli and involucres with mucilage containing schizogenous cavities; chloroplast 1 (-4) per cell with pyrenoid; antheridia (to 45) per cell chamber; spores grey, dark brown to blackish with a defined trilete mark; pseudoelaters thin or thick walled; capsule with stomata.

Order – Anthocerotales Limpr. *in* Cohn
Subclass – Notothylatidae Duff et al.

Thalli solid; chloroplast 1 (-3) per cell with or without pyrenoid; antheridia 2-4 per chamber with non-tiered jacket cell; sporophyte short, lying horizontally in the thallus, enclosed within involucre; stomata absent; columella present or absent; pseudoelaters absent or elongated with or without annular thickening; spore yellow to blackish with an equatorial girdle.

Order – Notothyladales Hyvönen & Piippo
Subclass – Dendrocerotidae Duff et al.

Epiphytic or epiphyllic; thalli solid or with mucilage containing schizogenous cavities; thalli with midrib, *Nostoc* present as bulging globose colonies in the ventral and dorsal side of the thallus; chloroplast 1-per cell with a conspicuous pyrenoid; antheridia 1-per chamber with non-tiered jacket arrangement; pseudoelaters with helical thickening, stomata absent; spores multicellular due to endosporic germination, pale yellow;

Order – Phymatocerales Duff et al., **Dendrocerotales**
Hässel emend. Duff et al.

BRYOPHYTA Schimp.

The characteristics of the major classes under the division Bryophyta (mosses) are given below –

Class – Takakiopsida Stech & W. Frey

Leaves divided into terete filaments; capsule dehiscent by single longitudinal spiral split.

Order – Takakiales Stech & W. Frey

Class – Sphagnopsida Ochyra

Leaves composed of a network of chlorophyllose and hyaline cells; branches usually in fascicles; capsule elevated on a pseudopodium; no setae; stomata lacking.

Order – Sphagnales Limpr., **Ambuchananiales** Seppelt & H.A. Crum

Class – Andreaeopsida Rothm.

Plants grow on acidic rocks; generally autoicious; calyptra small, capsule valvate with four valves attached at apex; seta absent, pseudopodium present.

Order – Andreaeales Limpr.

Class – Andreaeobryopsida Goffinet & W.R. Buck

Plants on calcareous rocks, dioecious; cauline central strand lacking; calyptrae large and covering whole capsule; capsules valvate, apex eroding and valves free when old; stomata lacking; seta present.

Order – Andreaeobryales B.M. Murray

Class – Oedipodiopsida Goffinet & W.R. Buck

Leaves unicostate; calyptrae cucullate; capsule symmetric and erect, neck very long; stomata lacking; capsules gymnostomous.

Order – Oedipodiales Goffinet & W.R. Buck

Class – Polytrichopsida Doweld

Plants typically robust; dioecious; cauline central strand present; leaves costate, often with ventral lamellae on costa; capsule operculate; seta present; peristome nematodontous.

Order – Polytrichales M. Fleisch

Class – Tetraphidopsida Goffinet & W.R. Buck

Leaves unicostate; calyptrae small conic; capsule symmetric and erect, neck short; peristome nematodontous, of four erect teeth.

Order – Tetraphidales M. Fleisch

Class – Bryopsida Rothm.

Plants small to robust; leaves costate; typically lacking lamellae; capsule operculate; peristome arthrodontous.

Subclass – **Buxbaumiidae** Ochrya

Leaves ecostate; calyptra cucullate or mitrate; capsule strongly asymmetric and horizontal, neck short; peristome double.

Diphysciidae Ochyra

Gametophore small, perennial; leaves costate, capsule immersed among long perichaetial leaves; peristome double.

Timmiidae Ochyra

Plants acrocarpous; leaves with sheathing base, costa single; laminal cells short, mammilose on upper surface; peristome double; calyptra cucullate.

Funariidae Ochyra

Plants terricolous (living on ground), acrocarpous, stem with central strand, annulus often well developed.

Dicranidae Doweld

Peristome haplolepidous, 4:2:3; late division in IPL asymmetric.

Bryidae Engl.

Peristome double of alternating teeth and segments; endostome ciliate, late stage division of IPL symmetric.

Chapter 3

Collection, preservation, working out the morphology

A. COLLECTION

Where and when to collect bryophytes - Though bryophytes are generally confined to humid habitats, avoiding direct sunlight, they are found in almost every ecosystem. In temperate and tropical rain forests, bryophytes, especially liverworts, form luxuriant epiphytic communities that play important ecological functions. According to Venderpoorten et al. (2010) habitats with lots of bryophyte cover should be looked into to find high species diversity. It is also suggested that collection must be done to verify field identification and to look for minute species hiding among the more prominent ones.

Suitable season for collection depends on the species and the location – (i) Generally bryophytes grow well after the rainy season, (ii) epiphytes grow well in winter, (iii) flood plain species are found once the water recedes, (iv) most of the sexual structures mature in spring or during rainy season.

How to collect – (i) Bryophytes may be collected in small tin, plastic bags or old paper envelops. Materials will be fresh for examination and may be used for further cultivation. (ii) Habitat notes should be recorded on the envelope along with date and season of collection. GPS data should be noted for exact location. (iii) Collected specimens should be spread, rapid air dried, attached soil should be cleared but colony formation should not be disturbed. Packets of newspaper are also preferred as it allows rapid water loss than paper bags, (iv) Epiphytic species should be collected with a shallow strip of the bark. Epiphylls should be collected with their underlying leaf. While collecting epiphytes – host species, type of bark, height on the tree and side aspects of the tree (N-S-E-W) should be noted, (v) Tiny bryophytes when collected, its extraction can be accomplished with the aid of masking tape.

What to record in field – (i) Starting collection number. When you leave the site finish note book recording with last collection number, (ii) Date, (iii) Location, (iv) Habitat – specific habitat and microhabitat, elevation, substrate exposure, indication of moisture, (v) GPS coordinates, (vi) Colour, growth form and fertility, (vii) Tentative identification in the field.

Amount of sample to be collected - The amount of sample to be collected depends on the following points – (i) Some samples to be deposited in the herbarium,

(ii) Some samples for working out the morphology, (iii) Some sample to be sent to some other worker, (iv) Small amount of samples to be collected in case of a rare species. Conservation must be a consideration, (v) Species occurring together should be noted on a packet.

How to organize sample plots

(i) Following Van der Poorten et al. (2010): Sampling methodology should properly be selected to understand patterns of community and taxon diversity at the landscape scale. Collection of plant species data using plots is commonly used. The bounded nature of plots in relation to a specific sample area allows for quantitative sampling of species abundance and frequency and the data will be used for statistical analysis. This plot method has been used for studying population and community dynamics in bryophytes. Plots may be organized in a regular fashion using a systematic grid or selected at random. There may be a systematic grid of 10 x 10 km with in which 'standard releves' of 100 m^2 are inventoried. In each releve all bryophyte species are collected and determined and voucher specimens are kept. This approach is most appropriate to identify the commonest species and assess their frequency and distribution. As many bryophyte species exhibit high specificity to peculiar meso- and microhabitat conditions.

(ii) Following Papp et al. (2005): The method involved for sampling bryophyte assemblages are as follows – (a) list of species weighted with abundance, (b) systematic sampling on the ground level (including soil, rocks, decaying wood) recording presence /absence data, (c) sampling of epiphytic bryophyte vegetation (only in forests).

The size of the permanent quadrates varies according to the nature of habitat to be studied. 10 m X 10 m in wetland, dry grass land and alkaline areas. 16 m X 16 m is the quadrate size in forests. Where the quadrate is 10 m X 10 m (100m^2) in size the bryophyte vegetation is systematically sampled in 25 sampling units of 0.5 X 0.5 m size that are set out on a grid at 2m interval. Where the quadrate size is 16 m X 16 m and the number of sampling unit is 64 using the same grid design.

In the sampling units the occurrence of species are recorded (presence/ absence data) and the substrates (soil, rock, decaying wood) of the species are marked. Depending on the sampling procedure following variables are calculated – (a) species richness of the quadrate, (b) mean species richness of sampling unit, (c) Simpson diversity and evenness of quadrates, (d) frequency and relative frequency of species.

B. PRESERVATION

i. Hepatics are best preserved in FAA (Formalin Acetic Alcohol). This is prepared by mixing 90 ml. Ethyl alcohol, 5 ml. Glacial acetic acid and 5 ml. Formalin. Subsequently materials may be stored in 70% alcohol.

Mosses are best preserved by drying them under light pressure within the folds of a piece of blotting paper. Quick drying is essential as it not only removes moisture effectively but also helps both the hepatics and the mosses to retain its original colour. Dry mosses often provide many important diagnostic characters – such as the characteristic mode of curling of the leaves in many mosses, character of the capsule, curling of peristome teeth. These features are utilized in their identification.

ii. Herbarium sheets may be prepared along the same line as those followed for the flowering plants. Lightly pressed dried specimens can be mounted directly on standard size herbarium sheet (42 cm × 29 cm).

iii. Alternatively paper or cellophane folders can be used. Specimens are put into the folders which may in turn may be mounted on a sheet or kept in a box in the form of a card index. Name of locality and habitat detail to be put on the envelope.

iv. Preparation of permanent slide: This procedure is useful in the preservartion of small liverworts or mosses for future critical study. The steps are as follows-

   a. Glycerin jelly preparation: Stir 10 gm gelatin in 60 ml of distilled water until dissolved. Add 70 ml. glycerin and 1 ml. phenol. Stir with gentle heat until dissolved. Label with date and initial and store in refrigerator. The material is mounted in glycerin jelly. After few days the material is to be remounted in glycerin jelly. The material covered by circular coverslip and sealed.

   b. Gum chloral preparation: Dissolve 40 gm Gum Arabic in 100 ml. cold distilled water. Then add 20 c.c. of conc. Glycerine and 50 gm. Chloral hydrate. Heat gently in water bath until the chloral hydrate is dissolved. Then filter through Whatman filter paper. Transfer the material direct to gum-chloral making the drop big enough to spread as a ring around the cover-slip.

C. WORKING OUT THE MORPHOLOGY

1. To study the internal structure of thallii of liverworts or hornworts

   a. Transverse sections are cut with blade.

   b. A thin section is passed through alcohol grades and stained in Bismark Brown following the steps mentioned below –

(i)   Thin section is passed through 30%, 40% and 50% alcohol with a forceps. Sections are kept in each alcohol grade for 5 minutes.

(ii) The sections are then put in Bismark Brown dissolved in 50% alcohol.

(iii) After staining for 6-8 minutes, the sections are washed in 50% alcohol to remove excess stain.

(iv) The sections are then passed through alcohol grades of 60%, 70%, 80%, 90% and absolute alcohol (5-6 minutes in each grade) to complete the dehydration process.

(v)   Then the sections are put in a 1:1 mixture of Absolute alcohol and Xylol and subsequently in pure Xylol.

(vi) Finally a drop of Canada balsam or DPX mounting medium is taken on a slide and the section is placed on it. The mounting medium with the section is covered by coverslip and dried on a hot plate.

2. To study a moss plant

a. The dried material should be soaked in water in a Petridish before working out the detail morphology for description and taxonomic identification.

b. A few leaves of moss gametophore may be detached carefully, dissecting them off the stem into drop of water that has been placed in the middle of a slide. The whole process should be conducted under a dissecting microscope. The leaf is covered by a coverslip and studied under microscope. Leaf may be passed through alcohol grades (as mentioned earlier) for dehydration and stained with Bismark Brown before mounting on Canada Balsam to make a permanent preparation.

c. A thoroughly soaked moss capsule should be cut off above the neck and then cut longitudinally in half on a slide in a drop of water. With needles (under a dissecting microscope) the spores may be removed and the delicate inner peristome may often be detached completely from the rim of the capsule mouth. This gives a satisfactory mount consisting of – (i) two fragments of capsule wall with outer peristome teeth, (ii) two detached piece of inner peristome. Then they are observed under low and high power of the compound microscope.

d. Mouth of the capsule may be cut in the form of a ring and then straightened to observe the peristome teeth.

e. Spores taken out from the capsule can be studied under Scanning Electron Microscope to reveal the ornamentation pattern of the outer wall which is of much taxonomic significance.

f. Study of spores of mosses : Morphological parameters of spores of mosses are important taxonomic characters. Detail ornamentation of spore external wall should be studied with light microscope (LM) and Scanning electron microscope (SEM). Spore material used for study can be taken from herbarium or can be collected freshly. Acetolysis method (Erdtman, 1960) should be employed for acetolysis and acetolysed spores are taken for measuring different parameters under light microscope.

Acetolysis method

i. The spores are centrifuged at 3000 r.p.m in distilled water for 5 min.

ii. The water is decanted.

iii. Glacial acetic acid is added and kept for 10 min.

iv. Spores are centrifuged for a further 5 min.

v. Glacial acetic acid is decanted off.

vi. Acetolysis mixture (acetic anhydride and conc. H_2SO_4 mixed in a ratio of 9:1) is added.

vii. The spores are heated in a water bath at 70°C for 5 min.

viii. The acetolysis mixture is decanted off.

ix. Again glacial acetic acid is added and centrifuged for 5 minutes and decanted off.

x. Finally distilled water is added, centrifuged for 5 minutes and decanted off.

xi. The acetolysed spores are treated with serial grade of alcohol (40% to absolute alcohol) for dehydration and finally stored in absolute alcohol.

Untreated spores should be prepared with glycerine-jelly onto microscope slides.

Preparation of glycerin-jelly – (as given in page 15)

At the time of study, place container in hot water to melt gelatin. Put one drop on slide, dust spore on it and place cover slip on it. For a semi-permanent slide, paint fingernail polish around cover slip or melted paraffin wax.

Next, measurement of the shortest and the longest diameters (in polar view) and the length of the polar axis and equatorial diameter (in equatorial view) should be taken in 25 randomly selected spores.

Polar axis – The imaginary line between the proximal and distal pole of the spore is called the polar axis (PA) which passes through the centre of the spore to the centre of the tetrad.

Equatorial diameter – The plane perpendicular to the polar axis through the middle of the spore is the equatorial plane - equatorial diameter (ED).

For SEM investigations, the unacetolyzed spores are directly placed on stubs, sputter-coated with gold and examined with SEM. Spore ornamentation pattern will be studied by SEM.

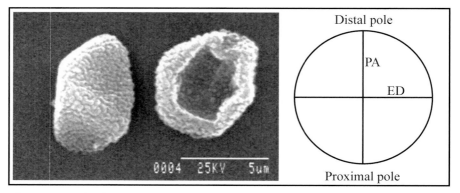

Fig. 3.1. (a) Distal and proximal end of a spore (b) Diagrammatic representation of polar axis and equatorial diameter

SPECIES FACT SHEET

Common name:	
Scientific name:	
Taxonomic note:	
Technical description:	
Distinctive characteristics:	
Similar species:	
Other descriptions and illustrations:	
Range distribution and abundance:	
Habitat association:	
Threats:	
Conservation considerations:	
Name of the worker and date of collection:	
Photo attachments:	

Chapter 4

General Morphology of Liverworts, Hornworts and Mosses

A. Thallose Liverworts (Marchantiopsida)

General features for description of genera

- Size of the thallus – plants robust or small delicate
- Branching pattern of the thallus
- Colour of the thallus and the margin
- Nature of the margin
- Internal organization of the gametophytic thallus – presence or absence of air pore, air chamber, organization of photosynthetic region
- Nature and arrangement of scales
- Shape of gemma cup, if present
- Extent of development of antheridiophore and archegoniophore; shape of male and female receptacle
- Position of antheridiophore and archegoniophore – dorsal, apical, ventral
- Capsule embedded in thallus or in involucre, not embedded
- Shape of involucre; presence or absence of perianth
- Structure of sporophyte, number of sporophyte per involucre
- Elater morphology
- Spore ornamentation

Description and systematic position of some commonly occurring genera

Division – Marchantiophyta

        Class – Marchantiopsida

                Subclass – Marchantiidae

                        Order – Lunulariales

                                Family – Lunulariaceae H. Klinggr.

                                      Genus – *Lunularia* Adans.

Habitat: Prefers damp shady places, at the base of damp wall, also on shady soil, found on rocks and grows as weeds near garden.

Gametophytic thallus pale green, irregularly furcated or innovating at the apex, large patches of overlapping individuals; shiny surface dotted with tiny pores (areolate); dorsal photosynthetic layer narrow, pore simple, air-chambers in one-layer, with filaments, not partitioned; ventral side with rhizoids and unicellular scales; semi-lunar (crescent) shaped gemma-cups on the dorsal surface; plants dioecious; male receptacles disciform, sessile at the apex of a short branch surrounded by elevated boarder of the thallus; female receptacle with stalk at the terminal end of a short branch, receptacle disciform composed of 4-cruciate tubular involucres, each containing one sporophyte; without rhizoid furrow, perianth absent; capsule rare with long pedicel, cells of capsule wall without annular band, dehiscing nearly to the base by 4 narrow valves; elaters bispiral, thread like.

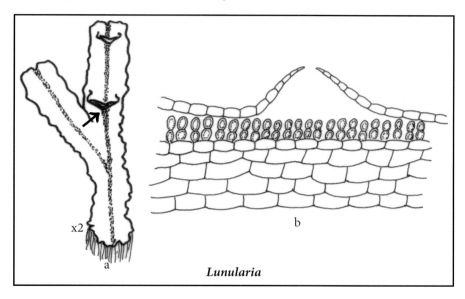

Fig. 4.1. (a) Thallus with crescent shaped gemma cup (→) (b) T.S. of thallus showing air pore and photosynthetic zone

Order – Marchantiales
Family – Marchantiaceae Lindl
Genus – *Marchantia* L.

Habitat: Species grow on soil or sometimes on rocks in very wet shaded ares. Grow best in sub calcareous soil condition under full sunlight and form mat.

Gametophytic thallus dichotomously branched, flat, broad midrib present, not translucent, hexagonal areolae on the dorsal side; dorsal photosynthetic zone narrow, air-chambers partitioned and in one layer, photosynthetic filaments within air-chambers are branched, air-pores barrel shaped; on the ventral side scales are in two or three rows on each side of the mid-rib, inner scale large and appendiculate, middle one small without appendage; tuberculate and smooth walled rhizoids present; tuberculate rhizoids arise in strands from the midrib also given off from the surface of scales; goblet shaped gemma-cup present on the dorsal side of the thallus, gemmae more or less biconvex, vertically inserted; plants dioecious; antheridiophore long peduncled, disciform, stelately or palmately lobed; archegoniophore with long stalk having air-chambers on the posterior side, receptacle stellate with 4 – 10 elongated rays; involucres 2-valved, alternating with the rays, enclosing several capsules, each surrounded by a perianth; male and female receptacle devoid of bristle; capsule with well developed seta, dehiscing to below the middle by irregular valves, capsule wall one layer; elaters simple, long, bi-spiral; spores small, tetrahedral.

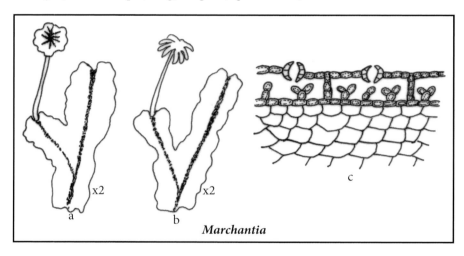

Fig. 4.2. (a) Thallus with antheridiophore (b) Thallus with archegoniophore (c) T.S. of thallus showing air pore

Family – Aytoniaceae Cavers

Genus –*Asterella* P. Beauv.

Habitat: Grows generally in moist places, sometimes on dry rocks.

Gametophytic thallus foetid (smelling like fish), prostrate, green, small or medium, simple, rarely divided, innovating from the apex; dorsal photosynthetic layer low, chambers narrow, often very irregular, in one or several layers, empty; pores simple, slightly convex; ventral scales with appendages, in one row on each side of the midrib; plants monoecious or dioecious; male receptacle sessile, naked, disc shaped or cushion like, just behind the stalk of the female receptacle; female receptacle terminal, on the main thallus, stalked, stalk with one rhizoid furrow covered with scales; receptacle flat, convex, conical or umbonate, usually 4 lobed; involucres arising from the margin of the lobes; archegonium one in each involucres; perianth usually ovate or oblong with an obconic apex;, reddish tinge on stalk and pseudo-perianth; capsule globose, shortly pedicelled; elaters short, simple or furcated; spores tetrahedral.

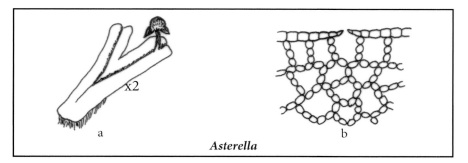

a   *Asterella*   b

Fig. 4.3. (a) Thallus with archegoniophore (b) T.S. of thallus showing air chambers in several layers

Family – Aytoniaceae Cavers

Genus –*Plagiochasma* Lehm. & Lindenb.

Habitat: Grow in dense patches on moist exposed wall (partly shaded) and crevices of rocky area.

Gametophytic thallus, green, not foetid, linear, broad, caespitose; dorsal side has narrow air-chambers , in several layers, empty pores sometimes with thick radial walls of the cells bounding them; ventral surface purple coloured specially at the margin; ventral scales in two rows, with single rounded or ovate appendages; plants monoecious; male receptacle sessile, usually horse-shoe shaped, surrounded by linear scale; air-chambers with simple pores between antheridia; female receptacle sessile when young, usually stalked at maturity; stalk arising from the dorsal side of the thallus, without a rhizoid furrow, with scales at base and apex, receptacle concave at the dorsal surface, 2 – 9 lobed; involucres large, inflated, bivalve; capsule short with a large foot, opening by an indistinct lid; elaters small, bi- or tri-spiral; spores large, yellow, tetrahedral.

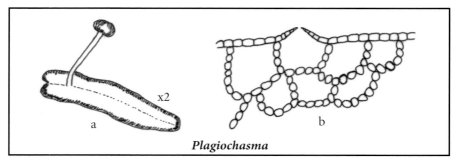

Plagiochasma

Fig. 4.4. (a) Thallus with archegoniophore (b) T.s. of thallus showing air chambers in many layers

Family – Ricciaceae Rchb.

Genus – *Riccia* L.

Habitat: Common in moist places, most are terrestrial, one aquatic species *R. fluitans*, found on clay, loam, in the fields, clay banks, on moist woodlands.

Gametophytic thallus prostrate, dorsiventral, dichotomously branched, several dichotomously branched thalli form a rosette form; each dichotomy is linear to wedge shaped, median region thickened with conspicuous longitudinal furrow on the dorsal side; ventral surface has transverse row of scales, crowded near the apex; in addition to scales there are two types of rhizoids – smooth walled and tuberculate (with internal peg like projections of the wall); internally dorsal side has compact green assimilatory region with vertical rows of chlorophyll bearing cells separated by narrow vertical air canals (no definite air-chamber); each air canal open to the outside by simle air pore; ventral side has the storage region with colourless parenchymatous tissue; most species are monoecious, few are dioecious; in all the species rounded antheridia and flask shaped archegonia are sunk within the tissues of the thallus; the capsule remains embedded in the thallus, contains no sterile cells, only spores which are large and few in number; older parts of the plants decay to release the large black spores.

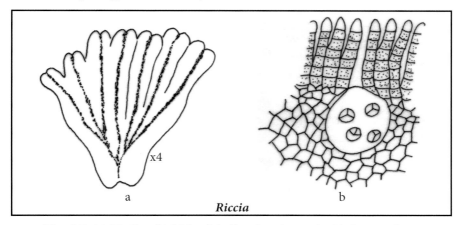

Fig. 4.5. (a) Thallus (b) V.S. of thallus showing embedded sporophyte

Family – Targioniaceae Dumort.

Genus – *Targionia* L.

Habitat: Grow in moist and shady places to form deep green overlapping individuals which are fixed to the soil only at their base.

Gametophytic thallus prostrate with ventral innovations near the apex or nearly dichotomously divided; dorsal surface with distinct areolae, pores simple, air-chambers distinct, full of filaments; ventral scales in two rows; plants monoecious or dioecious; antheridia on the dorsal surface of the disc like ends of short ventral innovations arising from the midrib or in long and broad mid-dorsal cushions on the main shoots; archegonia several, sporogonium usually one in each archegonium;

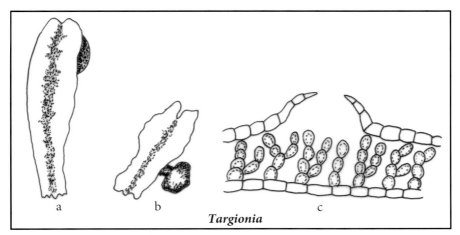

Fig. 4.6. (a) Dorsal view of a female thallus (b) Thallus showing male shoot (c) T.S. showing air pore and photosynthetic zone

capsule shortly pedicillate, with a well developed foot; involucres 2 valved, perianth absent; elaters long, bispiral, spores reticulate.

Family – Dumortiaceae D. G. Long

Genus –*Dumortiera* Nees

Habitat: Grows on moist earth or actually under running water on stones, on the walls of water reservoirs in rather dark places.

Gametophytic thallus very large, green, overlapping, translucent, repeatedly dichotomously branched or branching by apical innovations, apex notched, margin undulate, mid-rib very prominent, air-chambers absent, cells of outermost layer contain chlorophyll; no differentiation of photosynthetic and storage region scales on the ventral side greatly reduced, hyaline, attached on each side of the midrib in one row; gemma-cup absent; plants monoecious or dioecious; male receptacle disciform, depressed in the center, sessile with bristle in the margin; female receptacle sessile when young, stalked when mature, stalk long, covered with scales at the top, receptacle disciform, with few bristle like hair; 6 -10 short lobes, horizontal involucres; perianth absent; capsule with short pedicel, dehiscing

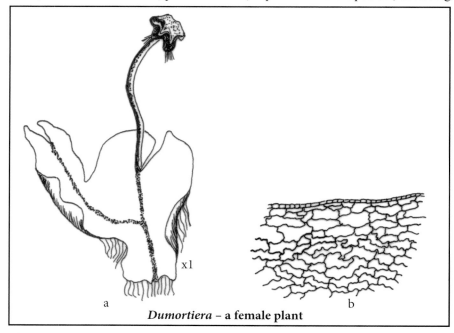

Dumortiera – **a female plant**

Fig. 4.7. (a) Thallus with archegoniophore (b) T.s. of thallus showing absense of air pore and chamber

in 4-8 valves, with elater like cells arising from bottom, elaters long, 2-4 spiral; spores warty tetrahedral.

Identification of some commonly occurring genera of Marchantiopsida

1a	Plants usually without air-chambers, pores and ventral scales	*Dumortiera*
1b	Plants with air-chambers, pores and with scales on the ventral surface	2
2a	Capsule embedded in thallus	*Riccia*
2b	Capsule not embedded in thallus	3
3a	Plants with gemma cup	4
3b	Plants without gemma cup	5
4a	Gemma cup goblet shaped	*Marchantia*
4b	Gemma cup crescent shaped	*Lunularia*
5a	Sporophyte in a tubular, bivalved involucres just under apex	6
5b	Sporophyte apical or dorsal in stalked receptacle	7
6a	Plants robust, large; involucre purple; scales conspicuous	*Targionia*
6b	Plants delicate, small; involucre hyaline; scales minute	*Cyathodium*
7a	Both male and female receptacle stalked	*Pressia*
7b	Male receptacle sessile	8
8a	Female receptacle distinctly dorsal; male receptacles horse-shoe shaped	*Plagiochasma*
8b	Female receptacle terminal, marginal or in the fork between two lobes; male receptacle cushion shaped, plants strongly foetid	*Asterella*

B. Simple thalloid and Leafy liverworts (Jungermanniopsida)

Order: Jungermanniales – Leaves succubous, incubous or transverse, undivided or variously lobed; usually with smaller lobes or lobules, dorsal, rarely with inflated water sac; under leaves present or absent; rhizoids fascicled from underleaves base or scattered along the ventral side of the stem; branches exogenous or endogenous; spore germination exosporic.

General features for description of genera

- Gametophyte - Thalloid or foliose.
- Plants flat or tubular ribbon like.
- Nature of branching of thallus – pinnately, monopodial.

- If thalloid – whether differentiated into midrib and wing; internal tissue organization undifferentiated.
- Presence or absence of hairs on ventral surface.
- If leafy – whether clearly differentiated into stem like axis and leaves.
- Plants monoecious or dioecious.
- Archegonia / Sporophyte anacrogynous or acrogynous.
- Position of antheridia and archegonia – on lateral branches.
- Presence or absence of gemmae.
  - Position of gemmae – on the ventral surface, or on dorsal margin.
- Leaves anisophyllous or isophyllous.
  - whether arranged in 3 equal rows or not arranged in three equal rows.
  - Shape of under leaves – triangular, oblong, rectangular; number of lobes in leaves.
  - Leaf arrangement - succubous, incubous or transversely attached.
  - Leaves distinctly bilobed or complicatedly lobed.
  - Leaf margin – spinose-dentate, ciliate-dentate; number of teeth per leaf, leaf apex – acute, acuminate, obtuse.
  - Presence or absence of vitta like cells.
- Stem cells as revealed in transverse section – whether differentiated into cortical cells and medullary cells.
  - Stem – leptodermous or not leptodermous.
- Nature of oil-bodies.
- Combination of these characters

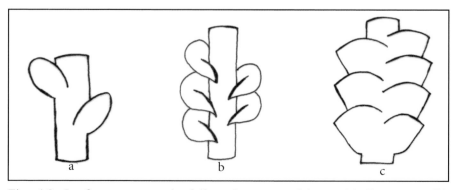

Fig. 4.8. Leaf arrangement in foliose Jungermanniales – (a) Transverse (b) Succubous (c) Incubous

Description and systematic position of some commonly occurring genera - Simple thalloid genera

Class – Jungermanniopsida
Subclass – Pellidae
Order – Pelliales
Family – Pelliaceae H. Klinggr.
Genus – *Pellia* Raddi

Habitat: Grow on moist soil or on rocks by the side of ditches or springs, also grow as patches among mosses or grasses.

Gametophytic thallus is thin, dorsiventral, prostrate, dichotomously branched with wavy margin; median midrib present, each branch terminates in a median notch; there is no internal tissue differentiation, some midrib cells are elongated longitudinally for conduction; ventral surface with smooth unicellular rhizoids, mucilage hairs present at the apex; plants are monoecious or dioecious; male plants bear antheridia all along the dorsal face of the midrib; archegonia develop anacrogynously in clusters just behind the growing tip of female plant; in monoecious plant the antheridia are behind the archegonia; within each involucre only one sporophyte develops, at maturity it has a long seta; capsule wall shows four vertical bands of thin walled cells; elaterophore present; spores germinate while already encased within capsule.

Order – Fossombroniales
Family – Fossombroniaceae Hazsl.
Genus – *Fossombronia* Raddi

Habitat: Grow on wide range of habitats, on acidic open ground, stubble fields, tracks, bare patches in pasture.

Gametophytic thallus is prostrate, branched with folds and lobes; each branch shows distinct stem with two lateral rows of leaf-like appendages; multicellular mucilage hairs protect the tip; smooth walled rhizoids on ventral surface, deeply pigmented; superficial antheridia and flask-shaped archegonia develop on the dorsal surface of the stem in the anacrogynous manner; the archegonia are enclosed in a large campanulate involucres; mature sporophyte resembles *Pellia* ; capsule spheroid, dehiscence not valvate, without elaterophore; spore germination exosporic.

Fig. 4.9. (a) *Pellia* thallus with sporophyte (→) (b) *Fossombronia* – thallus showing sporophyte (→) on gametophyte

Subclass – Jungermaniidae

Order – Porellales

Family – Porellaceae Cavers

Genus – *Porella* L.

Habitat: Grow as a green patch on shaded moist rocks, trees and even on soil.

Gametophytic thallus is flat, dorsiventral with bi- or tripinnately branched leafy axis; three rows of leaves – two lateral on dorsal side and one small on the ventral side called amphigastria. Each leaf is simple, with one layer of cell, lateral leaves are bilobed: upper antical lobe larger, usually ovate, lower posterior postical lobe is smaller, lobes visible from ventral side, leaf lobules lingulate, arrangement of leaves is incubous; stem shows little differentiation of tissue, cortical cells are smaller with thicker wall, central cells larger; rhizoids arise from the basal portion of amphigastria; plants dioecious; antheridia are borne singly in the axils of densely overlapping bract like leaves on compact branches; archegonia develop acrogynously on short lateral branches of female plant; inflated perianth surround the group of archegonia; mature sporophyte consists of a spherical capsule, a short seta and a poorly developed foot; perianth mouth not beaked; capsule wall 3-6 layered; dehiscence along four vertical lines; elaters present; spores undergo endogenous divisions.

Family – Lejeuneaceae Cavers.

Genus – *Lejeunea* Lib.

Habitat: Grow mainly on tree barks and on moist rocks.

Gametophytic thallus minute and delicate, branching pinnately; shoots are endowed with stems and leaves in three rows; mature flattened dorsiventral shoot shows two lateral rows of alternating leaves on the dorsal side and one row of ventral amphigastria (rounded); leaves are simple, each leaf with upper antical lobe and posterior postical lobe, postical lobe smaller, saccate, arrangement of leaves incubous; plants monoecious or dioecious; mature sporophyte differentiated into foot, seta and capsule; young sporophyte covered by a calyptra; elaters are with single spiral thickening.

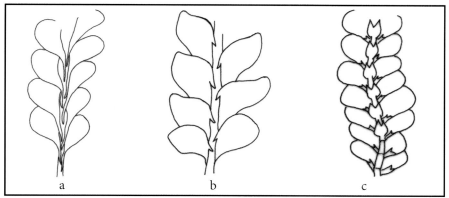

Fig. 4.10. Thallus showing leaf arrangement (a) *Plagiochila* (dorsal) (b) *Plagiochila* (ventral) (c) *Lejeunea*

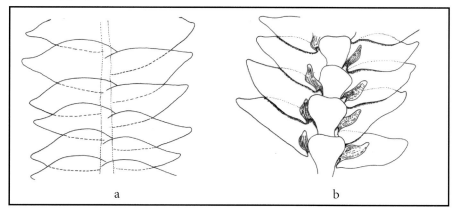

Fig. 4.11. Thallus showing leaf arrangement (a) *Porella* (dorsal) (b) *Porella* (ventral)

Order – Jungermanniales
Family – Plagiochilaceae
Genus – *Plagiochila* (Dumort.) Dumort.

Habitat: This liverwort forms green to dark green mats.

Gametophytic thallus large with toothed, unlobed (shortly bilobed in *P. hattorii*) decurrent leaves; underleaves (amphigastria) poorly developed, leaf margin spinose - dentate to ciliate – dentate; front margin is reflexed and extends down to stem; insertion line is oblique; leaves may be with or without vitta, leaves closely imbricate; branching terminal or intercalary; plants dioecious; perianth mouth wide; capsule wall polystratose; spore germination exosporic.

Identification of some commonly occurring genera of Jungermanniopsida (after Sing, D. et al. 2010)

1a	Plant body usually thalloid, sometimes leafy; when leafy, not clearly differentiated into stem and leaves; rhizoids scattered; sporophyte anacrogynous, dorsal, ventral or lateral in position, never in axil of leaf	2 (Metzgeriales)
1b	Plant body usually differentiated into stem and leaves; rhizoids scattered or fasciculate; sporophyte acrogynous, terminal on main or lateral branches, always in the axil of leaves	6 (Jungermanniales)
2a	Plant body leafy, not clearly differentiated into stem and leaves; leaves unistratose, alternate	*Fossombronia*
2b	Plant body thalloid	3
3a	Thallus always broad, light-pale green, often pigmented; mid-rib and hairs absent; male gametangia not defined, antheridia scattered on dorsal surface; archegonia strictly dorsal; capsules spherical, with basal elaterophore	*Pellia*
3b	Thallus broad, dark green, without midrib and hairs, or linear, light-pale green with midrib and hairs; both antheridia and archegonia on lateral or ventral branches; capsules ovoid, ellipsoid or cylindrical	4
4a	Thallus well differentiated into midrib and unistratose wing; hairs present; antheridia and archegonia borne on ventral branches	*Metzgeria*

4b   Thallus not differentiated into midrib and wing; hairs absent; antheridia and archegonia borne on lateral branches   **5**

5a   Thallus 2.5-5 mm. broad, simple – furcate; seta 8-12 cells across the diameter with numerous outer and inner cells   *Aneura*

5b   Thallus narrow, 0.3-1.5 mm. broad; profusely pinnately – palmately or irregularly branched; seta 4 cells across the diameter with 11-12 outer and 4-inner cells   *Riccardia*

6a   Leaves simple, not complicate bilobed, margins usually with prominent teeth   *Plagiochila*

6b   Leaves complicate bilobed, margins usually without teeth   **7**

7a   Leaves arranged in two rows; rhizoids borne in the center of the ventral surface of leaf lobule   *Radula*

7b   Leaves arranged in three rows; rhizoids usually borne at the base of underleaves ventral leaf or amphigastrium   *8*

8a   Leaf lobules saccate; hyaline papilla present on the lobule   *Lejeunea*

8b   Leaf lobules lingulate, lanceolate, galeate, cucullate, or explanate; hyaline papella absent   **9**

9a   Leaf lobules galeate, cucullate, or explanate; perianth mouth beaked; capsule wall 2-layered   *Frullania*

9b   Leaf lobules lingulate – lanceolate; perianth mouth not beaked; capsule wall 3–6-layerd   *Porella*

C. HORNWORTS (Anthocerotopsida)

General features for description of genera

- Nature of thallus dimension – with or without schizogenous cavities, spongy or not spongy, presence or absence of pyrenoid.
- Number of chloroplast per cell.
- Tiered or non-tiered jacket cell arrangement in antheridia
- Number of antheridia in antheridial chamber.
- Plants monoecious or dioecious.
- Distribution of stomata on capsule wall, number of stomata/sq. mm.
- Presence or absence of intermediate meristematic zone.

- Arrangement of pseudo-elaters within capsule and number of spiral thickening.
- Colour of spores – yellowish, dark brown to black.
- Ornamentation of sporoderm – spinose, reticulate, baculate, reticuloid, reticulations with spines, tubercles.
- Nature of proximal surface of spore – triradiate mark with papillae at the center or papillae uniformly distributed.
- Various combinations of the features mentioned above.

Identification of some commonly occurring genera
Description of some commonly occurring genera

Class - Anthocerotopsida
Subclass – Anthocerotoidae
Order – Anthocerotales
Family - Anthocerotaceae (Gray) Dumort. *corr.*
Trevis. *emend.* Hässel
Genus – *Anthoceros* L.

Habitat: Grow on very moist, shady places by clay banks and ditches or in the crevices of rocks.

Gametophytic thallus greasy dark green, simple, prostrate, dorsiventral, dichotomously lobed; no definite midrib; dorsal surface smooth or with ridges; ventral surface has only smooth walled rhizoids, scales and mucilage hairs absent; thallus and involucres with mucilage-containing schizogenous cavities, having *Nostoc* filaments; chloroplast 1 (-4) per cell; pyrenoid present; plants dioecious or monoecious; antheridia numerous (4-45) per chamber with a tiered jacket cell arrangement; archegonia embedded in the thallus; capsules with stomata, central columella surrounded by spores and short pseudoelaters, thin-walled; indeterminate meristematic zone well developed; spores smoky gray, dark brown to blackish with a defined trilete mark; ornamentation spinose, punctuate, baculate.

Subclass – Notothylatidae
Order – Notothyladales
Family – Notothyladaceae (Milde) Mull. Frib. ex Prosk.
Genus - Notothylas Sull. ex A. Gray

Habitat: Gametophytic thallus grows on shady moist soil or rock.

Gametophytic thallus is solid, yellow-green forming an orbicular or suborbicular rosette which has a characteristic pleated appearance; thallus anatomy resembles *Anthoceros*; it is 6 to 8 cells deep in the middle, thinning out to 1 to 3 cells in the edges; chloroplast 1 (-3) per cell, pyrenoid present or absent; Indian species are monoecious and protandrous; the antheridia and archegonia resemble those of *Anthoceros;* The mature sporophyte is only 2 to 3 mm long, tapering at both ends and lies horizontal on the thallus, scarcely projecting beyond the thallus margin; columella present or absent; it is usually completely enclosed by the thin, membranous involucre, massive sporogenous tissue (2-5 layers); the foot is much smaller than in *Anthoceros*, although haustorial outgrowths are well developed; the intermediate meristematic zone also is much less developed, the meristematic development being negligible so that the capsule does not grow much, the pseudoelaters are unicellular, of irregular form and have thickenings on the walls; sporophytes dehisce along one suture or absent; spores are often liberated by the decay of the capsule wall.

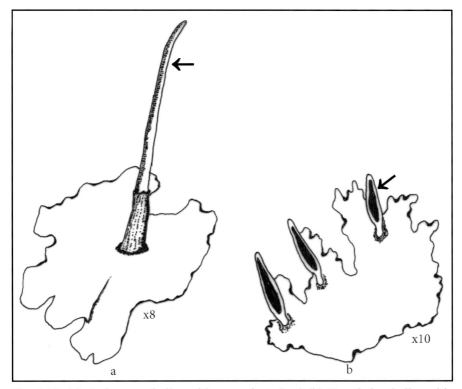

Fig. 4.12. (a) *Anthoceros* thallus with sporophyte (⟶) (b) *Notothylas* thallus with sporophyte (⟶)

Subclass – Dendrocerotideae
Order – Dendrocerotales
Family – Dendrocerotaceae (Milde) Hässel
Genus - *Dendroceros* Nees (tree horn)

Habitat: Grow on humid ground, rocks and on the sides of trees (as epiphytes).

Gametophytic thallus solid, is yellowish-green, thallus has a prominent ventral midrib and the edges of the thallus having a single layer of cells and have an undulating margin, symbiotic colonies of *Nostoc* grow as globose colonies in the ventral and dorsal side of the thallus; chloroplast (1-per cell), pyrenoid is spherical and contains irregularly-shaped subunits with electron-opaque inclusions; antheridia occur singly in rows along the midrib, non-tiered jacket cell arrangement; the sporophyte is erect when mature and may grow upto 5 centimeter tall, stomata absent, the interior of the sporophyte differentiates into a central columella and surrounded by spores and pseudoelaters with a helicoidal thickenings; divisions in the intercalary meristem persist longer than in *Notothylas* but not as long as in *Anthoceros*; dehiscence of the capsule would be by two longitudinal lines; spores are both green and relatively large with an ornamented surface; spores have chloroplast and germinate endosporically; multicellular spores.

Genus - *Megaceros* Campb. (long horn)

Habitat: Grow on rocks and soil banks in streams.

Gametophytic thallus is solid in rosette, of large size, have a branching thallus; chloroplasts in *Megaceros* occur in multiples (12 per internal thallus cell) and lack a pyrenoid; plants are monoecious; antheridia 1 (-2) per chamber with a non-tiered jacket cell arrangement; sporophyte without stomata and the spores are green because they contain chloroplasts; elaters are helical and with spiral thickening; central columella may contain 40 cells; massive foot of *Megaceros* contains thousands of small undifferentiated cells; spore green with chloroplast, have reticulate ornamentation.

Identification key of commonly occurring genera

1a   Plants with horizontally deflected sporophytes usually enclosed within the involucres   *Notothylas*

1b   Plants with erect, horn-like sporophytes projecting far beyond the involucres   2

2a   Thallus compact, without schizogenous cavities; spores yellowish with granulose-papillose sporoderm   *Phaeoceros*

2b   Thallus with schizogenous cavities; spores dark brown-black with spinose – baculate, reticulate-reticuloid sporoderm   3

3a   Proximal surface of spores devoid of distinct triradiate mark; sporoderm baculate; pseudo elaters usually 4-celled with highly thickened walls and dark, narrow lumen   *Folioceros*

3b   Proximal surface of spores with a distinct and bold triradiate mark; sporoderm spinose, reticulate or reticuloid; pseudo elaters usually 1-celled with thin walls and light, broad lumen   *Anthoceros*

D. MOSSES (Bryophyta)

General features for description of genera

The gametophytes of mosses are small, usually perennial plants, comprising branched or unbranched shoot system bearing spirally arranged leaves.

Growth form and archegonial position - Two common growth forms can be seen in mosses – acrocarpous and pleurocarpous species. **Acrocarpous mosses** are erect, upright shoot system that are branched or sparingly branched. Branching is sympodial with branches comparable to main shoot from which they arise. Main shoot grows separately or crowded together to form tufts or cushions. The archegonia, within perichaetia, and therefore sporophytes develop at the tip of the primary branch thus terminating growth along its axis. **Pleurocarpous mosses** are creeping plants lying flat on the substratum with extensive lateral branching and form mats. The indeterminant main stem may be morphologically distinct from the secondary and tertiary branches arising from it. The reproductive branch tip is pushed aside by lateral branches growing out below it. Archegonia do not terminate the main axis or axes of growth but develop at the tip of small lateral branches.

Cladocarpic mosses – Capsules that lack operculum and dehisce by breakdown of capsule wall. **Stegocarpic mosses** – Capsule with distinct operculum.

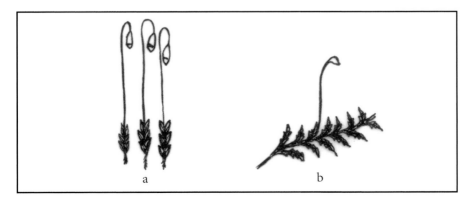

Fig. 4.13. (a) Acrocarpous moss (b) Pleurocampous moss

Detail morphological features –

- Leaves typically from all sides of the stem, most commonly exhibiting a spiral phyllotaxy. Distichous or tristichous arrangements can also be found.
- In a few mosses, mature leaves are clearly ranked; e.g., the leaves of *Fissidens* and *Bryoxiphium* are in two rows, a 1/2 phyllotaxy, and *Fontinalis* and *Tetraphis* have leaves aligned in three rows, a 1/3 phyllotaxy. In most mosses, however, the leaves are spirally distributed, with 2/5 and 3/8 phyllotaxies being most common (W. Frey 1971; B. Crandall-Stotler 1984). Mostly the leaves are isophyllous. Some are anisophyllous with either dorsal or ventral leaves decidedly smaller than the lateral leaves.
- Abaxial surface is defined as dorsal side and the adaxial surface as ventral side.
- Surface features may differ – one surface may be smooth and the other surface papillose.
- Moss leaves are generally lanceolate to ovate.
- Leaves are generally unistratose, only costa multistratose; in a few family leaf is multistratose.
- Margin of the leaf may bear a border of hyaline, elongated cells with thickened walls – such borders are called limbidia.
- Leaf margin may be plane, recurved, incurved and are often toothed.
- Laminal cells may be short and isodiametric (acrocarps) or these may be long and thin (pleurocarps).
- The basal cells of the leaf margin, known as alar cells are also frequently differentiated in size, shape, colour from other cells of the lamina.
- Laminal cells are smooth or variously ornamented, commonly bearing a variety of wall projections called papillae. Papillae may be solid or hollow, simple or branched, single or multiple.
- Costa may be percurrent (extend from base to shortly before its apex) or excurrent (costa going beyond its apex).
- Mosses may be monoecious or dioecious.
- In autoicous arrangement, there are separate androecia and gynoecia on the same plant, often on separate branches.
- Capsules vary in shape from spheroid to ovoid, obovoid, pyriform, ellipsoid or long cylindric.
- Capsule may be symmetric or asymmetric.
- Capsule can be erect, inclined, nodding or even pendant. They can be longitudinally furrowed.

- Capsule can be divided into three zones – the basal neck, median spore containing urn and the distal operculum.
- Neck of the capsule may taper gradually into the seta or it may be differentiated as a swollen apophysis.
- Stomatal complex present in exothecium of neck – may be superficial or sunken.
- The capsule opens by an apical pore, longitudinally split or opening by an operculum. The external cell layer (external to amphithecium – exothecium) often has stomata, especially in the neck. Underlying the exothecium there are parenchymatous cells. Both concentric layers constitute the amphithecium. Internally, the endothecium consists of a cylinder of sporogenous tissue, surrounding a columella of sterile cells. When the apex of the capsule falls off at maturity and reveals a structure called peristome. This is a ring of triangular segments surrounding the mouth of the capsule. Two basic types of peristomes are found in mosses. Peristome (a circular system of teeth that is inserted on the mouth of the urn) may be nematodontous (teeth constructed of bundles of whole, dead cells) or arthrodontous (teeth consisting of remnants of paired, periclinal cell walls).
- Nematodontous peristome teeth are structured by narrow columns of entire cell wall remnants. Each tooth consists of agglomerated cylinders formed by the periclinal and anticlinal thickened cell walls. Peristomial cells derived from the innermost amphithecium but multiple concentric peristomial layers (4-7) contribute to its formation.

  In arthrodontous peristome teeth three important rings of cells of the amphithecium are involved in the formation of teeth in most taxa. These three concentric rows are known as the 'outer', 'primary' and 'inner' peristomial layers (OPL, PPL, IPL). In arthrodontous mosses, each teeth is composed of periclinal cell wall remnants between two of the three concentric peristomial cell layers. If the teeth are formed by the tangential walls between the OPL and PPL, the row of teeth is collectively called exostome. In the second case, the cell wall remnants are located between the cell rings of the PPl and IPL, therefore the row of segments is known as endostome.
- Arthrodontous peristomes are of two types – **haplolepidous** (consisting of a single ring of 16 teeth that are formed by cell wall deposition on the paired wall of PPL and IPL). The outer surface of the tooth may be variously ornamented with horizontal striae, trabeculae or papillae. Teeth may be forked at the tip or fused at the base. In **diplolepidous** peristome, two sets of teeth are differentiated,

the exostome or outer peristome, formed by deposition on paired walls of OPL and PPL and the endostome formed at the PPL-IPL juncture. The architecture of endostome is variable with different patterns of surface ornamentation on outer and inner surfaces.

- In a diplolepidous – alternate peristome of the bryoid or hypnoid type, the endostome comprises a basal often keeled membrane, topped by 16 broad, perforate segments that alternate with exostome teeth.
- Peristomal formula – is the method to describe the number of cells in each of the three peristomial layers contributing to the peristome. For ease of comparison the peristome formula uses the number of cells in a 1/8 slice of a transverse section of capsule circumference. 4:2:4 formula means 32, 16, 32 cells in OPL, PPL, IPL.
- Spores are typically spheroid but may also be ovoid, reniform or tetrahedral.
- Ornamentation pattern –

Table: Some examples of spore exine ornamentations

Name of the Genus	Ornamentation types
Funaria	Baculate/Gemmate
Tortula	Scabrate-verrucate
Bryum	Dense piloid/blunt at the tip, sharp at the tip, hemispherical at the tip
Hypnum	Scattered piloid
Encalypta / Trematodon	Gemmoid
Bruchia	Elongate process type
Helicophyllum	Granulate
Syntrichia	Granular
Pterigoneurum	Verrucate
Cynodontium	Verrucate
Campylopodium	Convolute

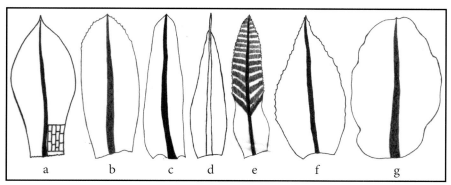

Fig. 4.14. (a) Apiculate (b) Mucronate (c) Ligulate (d) Long hair-point (e) Sheathing leaf base (f) Lanceolate (g) Obovate

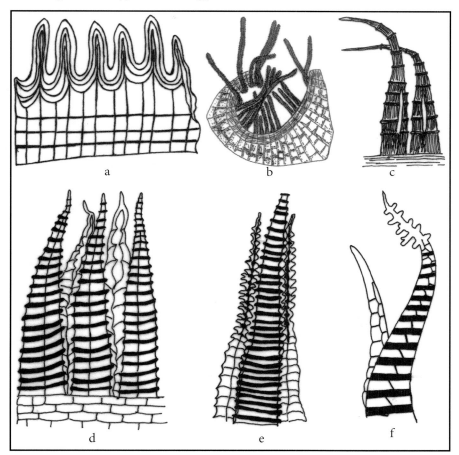

Fig. 4.15. Various types of peristome teeth (a) *Atrichum* (b)*Semibarbula* (c) *Trematodon*, (d) *Bryum* (e) *Mnium* (f) *Funaria*

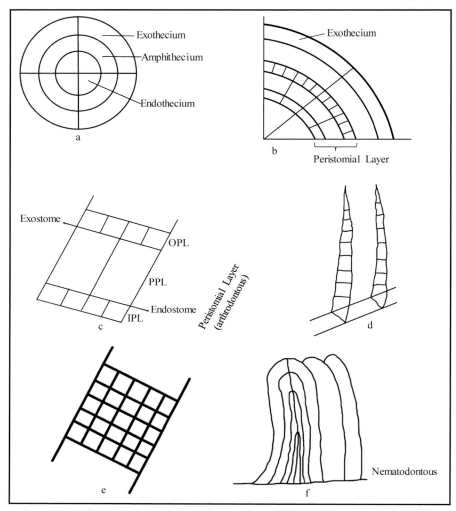

Fig. 4.16. (a) Zonation of developing sporophyte (b) Peristomial layers (c) Detail location of OPL, PPL, IPL for arthrodontous (d) Arthrodontous peristome teeth (e) Peristomial layers in development of nematodontous teeth (f) Nematodontous teeth

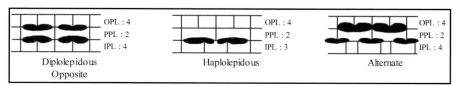

Fig. 4.17. Details of peristome formula

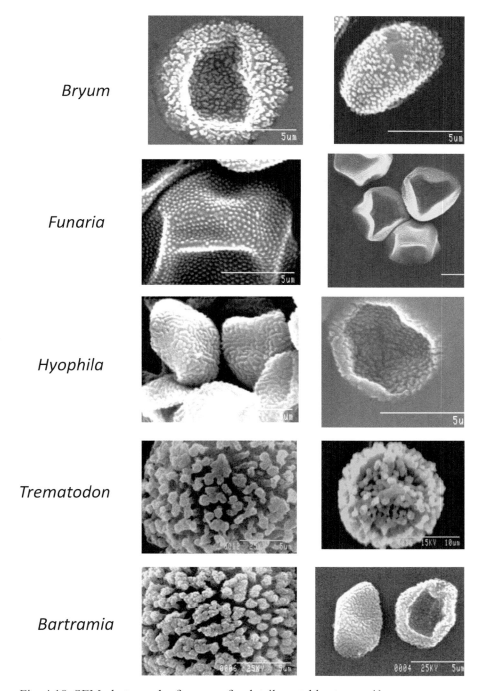

Fig. 4.18. SEM photograph of spores : for detail see table at page 41

Description of some commonly occurring genera

Genera arranged according to the classification proposed by Goffinet, Buck and Shaw (2009).

Class – Sphagnopsida

Order – Sphagnales

Family – Sphagnaceae Dumort.

Genus – *Sphagnum* L.

Habitat: Growing in very moist situations, usually in dripping water

Leafy gametophores robust, greenish-white to brown, long branching, each gametophore terminates in a dense tuft of apical branches called capitulum; leaves without costa, with cells one layer in thickness and with characteristic areolation; large rhomboidal, fibrose, porose, transparent cells bordered on all sides by narrow elongated, chlorophyllose cells which are narrow triangular in cros-section; stem leaves long, tongue shaped, branch leaves larger, long, elongate-oval, cymbiform; antheridia occur in short lateral branches near the apex of main axis, they are borne in the axils of perigonial leaves; archegonia are borne on short lateral branches at the cntre of capitulum; perichaetial leaves surround the archegonia; sporophyte elevated on pseudopodium (elongated gametophytic stalk), has a foot and capsule, connected by neck, covered by calyptra and leaves of gametophores, sporogenous tissue occupies a dome like position above the columella; spore with brown sculptured wall.

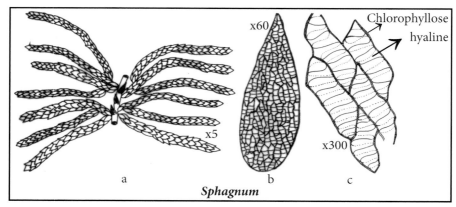

Fig. 4.19. (a) Gametophore with branches (b) A single leaf (c) Leaf enlarged to show two types of cells – chlorophyllose and hyaline

Class – Polytrichopsida
Order – Polytrichales
Family – Polytrichaceae Schwäger
Genus - *Atrichum* P. Beauv.

Habitat: Grows commonly in woods, also present on wate land in the open, can be found at rocky stream

Leafy gametophyte medium-sized to large, soft, yellowish green; lower leaves small, upper leaves erect, spreading, undulating, carinate, narrow lingulate from a non-sheathing base; lamina rough on back with teeth in oblique rows on both sides of the vein, costa narrow, toothed on back, ending in spines just below leaf tip; leaf-cells thick-walled, elongated rectangular at base, rounded hexagonal at top; capsule slightly bent, cylindrical, upto 8 mm. long; operculum conical, long, rostrate; calyptra narrow, cucullate; peristome teeth 32, nematodontous, with brown striolations.

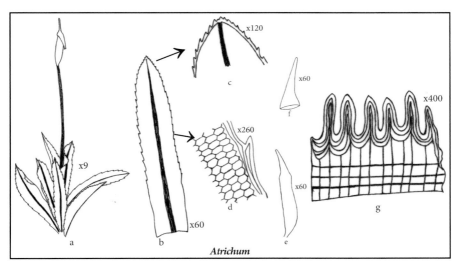

Fig. 4.20. (a) Gametophytic plant with sporophyte (b) Leaf, (c) Leaf apex (d) Alar cells (e) Capsule (f) Calyptra (g) Peristome teeth

Genus – *Pogonatum* P. Beauv.

Habitat: Grows commonly on rock soil in the hill stations

Leafy gametophyte in lax tufts; stem simple; lower leaves small, scaly, upper leaves rigid, erect-spreading, incurved when dry, lanceolate from a wider sheathing base, apex acute, margin toothed almost to the base of the lamina, costa reddish brown, strongly toothed on back, very wide in the lamina; longitudinal lamini numerous covering most of ventral face; topmost cells large, divided to base into two flask shaped cells; leaf base cells long, narrow rectangular, becoming shorter wider in lamina; capsule erect or inclined, reddish brown, ovate cylindrical; operculum convex, shortly rostrate, calyptra covers whole capsule; peristome teeth nematodontous, 32 with brown striations.

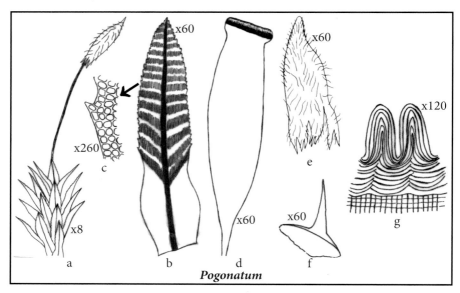

Pogonatum

Fig. 4.21. (a) Gametophyte with sporophyte (b) Leaf showing sheathing leaf base (c) Margin enlarged (d) Capsule (e) Calyptra (f) Operculum (g) Peristome teeth

Class – Bryopsida

Sub-class – Funariidae

Order – Funariales

Family – Funariaceae Schwäger

Genus – *Funaria* Hedw.

Habitat: Grows commonly on moist shady situations such as damp soil, shady banks, sometimes on the trunks of trees and on the wall.

Leafy gametophyte green to yellow-green, loosely or close tufted, simple or branched from base; lower leaves small, lax, showing poor development of costa; upper leaves large forming a rosette on top, oblong-obovate to oblong-lanceolate, concave, erect spreading, margin entire, apex acute, costa strong, percurrent or short excurrent in the upper leaves; cells of lamina thin walled, rectangular to subhexagonal, more elongated at base; capsule horizontal to pendulous, asymmetrical with an oblique narrower mouth, with apophysis, yellow with a deep red mouth; seta apical, erect, strongly arcuate; peristome diplolepidous, epicranoid, outer 16 spirally arranged, inner 16 hyaline, apices of teeth united to a small disc.

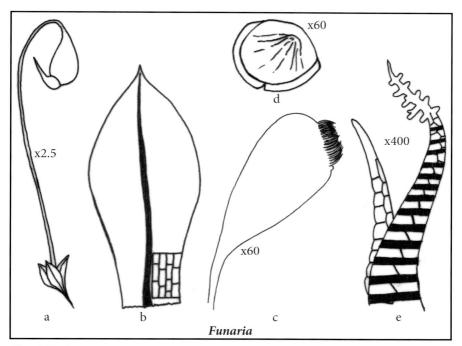

Fig. 4.22. (a) Gametophyte with sporophyte (b) Leaf with alar cells (c) Capsule (d) Operculum (e) Peristome teeth

Genus – *Physcomitrium* (Brid.) Brid.

Habitat: Grows commonly on soil, on some wet patch by road or ditch, on the bed of dried out pond, on garden soil, on the banks of river.

Leafy gametophyte small, caespitose, forming bright green patches; stem short, slender, erect; shoot with branch; lower leaves smaller, upper ones clustered, erect spreading in both wet and dry conditions, oblong ovate, acuminate, margin serrulate above; costa strong, percurrent in upper leaves, basal leaf cells thin-walled, rectangular; upper cells become narrower and shorter; seta slender, short; capsule globose with brightly coloured apiculate operculum; calyptra small, cap like with a narrow tip; peristome not developed; spores bright red-brown.

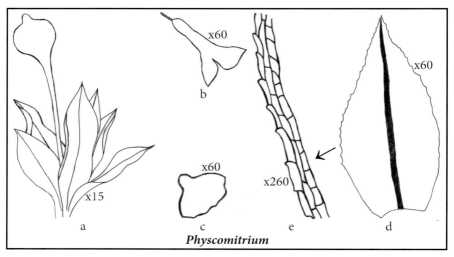

x60

b

x60

x60

x15

x60

x260

a   c   e   d

Physcomitrium

Fig. 4.23. (a) Gametophyte with sporophyte (b) Calyptra (c) Operculum (d) Leaf with costa (e) Alar cells

Order – Dicranales
Family - Bruchiaceae
Genus - Trematodon Michx.

Habitat: Grows commonly on sandy soil and rocks.

Leafy gametophyte yellow-green, grow in patches; leaves erect, spreading, flexuous when dry, 2.5-4.00 mm. long, tapering from a broad sheathing base to a long narrow canaliculated lamina, margin slightly toothed at apex, entire below; costa narrow at base, broader above, ending just below apex; leaf cells rectangular, becoming smaller rectangular above; capsule cylindrical, little curved, narrow cylindrical apophysis with swelling at base; calyptra circulate; peristome teeth red brown, haplolepidous, split slightly above base by longitudinal perforation but united at the tip, vertically striolate; spores with verrucate ornamentation.

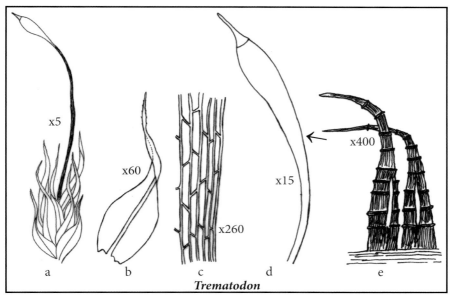

Fig. 4.24. (a) Gametophyte with sporophyte (b) Leaf (c) Alar cells (d) Capsule with long apophysis (e) Peristome teeth

Order – Pottiales
Family – Pottiaceae Schimp.
Genus – *Hyophila* Brid.

Habitat: Grows as dense tufts of dark green plants on bricks, rocks

Leafy gametophyte erect, simple or branched shoots; top leaves spreading in rosettes; leaves erect spreading, curled circinately with the leaf margins inrolled when dry, leaves oblong-lingulate, carinate, apex broadly pointed, margin entire below, denticulate at apex; costa red-brown, prominent, wider at base, percurrent; leaf base cells rectangular, smooth, becoming smaller above; seta apical, erect; capsule cylindrical, brown, erect, operculum conical, rosette, calyptra cucullate; peristome not developed.

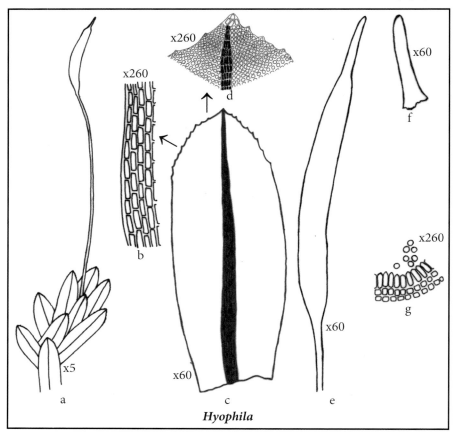

Fig. 4.25. (a) Gametophyte with sporophyte (b) Alar cells (c) Leaf (d) Leaf apex
(e) Capsule (f) Calyptra (g) Undeveloped peristome

Genus – *Semibarbula* Herzog ex Hilp.

Habitat : Grows commonly in dense tufts on old walls of brick and mortar (calciocole)
Leafy gametophyte, yellow-green to green, short, show luxuriant growth;
leaves lax, spiral, oblong to ovate-lanceolate, broad at slightly wider base; incurved
and curled when dry, margin papillose; apex blunt, leaf base cells elongated
rectangular; costa strong, light greenish yellow, percurrent; seta apical, reddish on
top, erect; capsule reddish brown, erect, cylindrical; operculum conic, short rostrate;
calyptras cucullate covering tip but reaching the middle of the urn; brown peristome
split into 32 short filamentous segments covered with dense minute papillose all
over the surface.

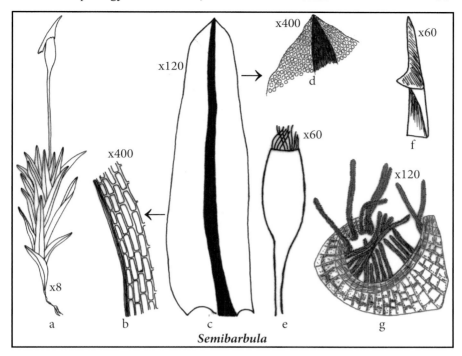

Fig. 4.26. (a) Gametophyte with sporophyte (b) Alar cells (c) Leaf (d) Leaf apex
(e) Capsule (f) Calyptra (g) Peristome teeth

Sub-class – Bryidae

Order – Bryales

Family – Bryaceae

Genus – *Bryum* Hedw.

Habitat: Grows densely on damp walls or calcareous soil, dry wall top, bare ground, marsh habitat, and mountain ledges.

Leafy gametophyte densely tufted, slender, bright to dull green; lower leaves smaller, upper leaves ovate to oblong lanceolate, erect, long acuminate, margin entire; costa strong, excurrent; upper leaf cells narrow, rhomboid to hexagonal, basal cells shorter; capsule red to purple, pendulous, spongy apophysis; capsule mouth wide, operculum big conical; peristome diplolepidous, metacranoid, transparent yellow, reddish, outer teeth broad, lanceolate with sharp, hyaline papillose tips.

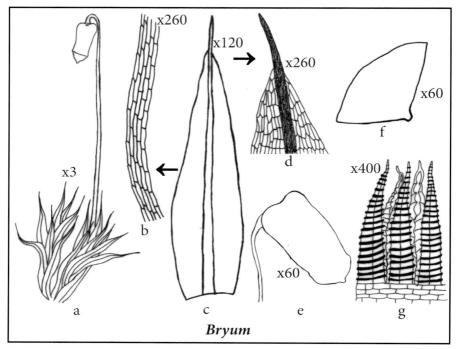

Fig. 4.27. (a) Gametophyte with sporophyte (b) Alar Cells (c) Leaf (d) Leaf apex (e) Capsule (f) Operculum (g) Peristome teeth

Genus - *Mnium* Hedw.

Habitat: Grows commonly on humus, peat, rotten wood and rock ledges at high altitude, form extensive carpets on the ground in woodland.

Leafy gametophyte yellow-green to dark green forming loose or compact creeping mats; leaves are fertile, erect shoot more crowded at apex, spreading, leaf, usually oblong, usually notched at tip, suddenly narrowed at base, undulated; margin flat strongly boardered, with teeth in one row almost to the base; sterile shoots almost complanate with leaves in two rows; costa red, percurrent or excurrent in the apiculus; leaf cells thick walled, rectangular and subrectangular at base; capsule horizontal to pendulous, yellow to light brown, ovate-oblong – cylindrical with short apophysis; operculum long, rostrate; peristome bryoid (as in *Bryum*).

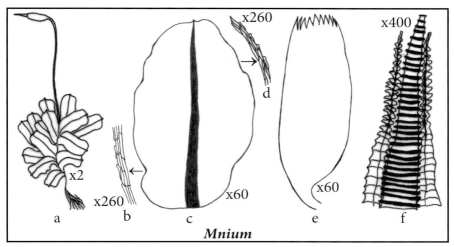

x260
x400
d
x2
x260
x60
x60
a   b   c   e   f
Mnium

Fig. 4.28. (a) Gametophyte with sporophyte (b) Alar cells (c) Leaf (d) Cells at upper margin (e) Capsule (f) Peristome teeth

Genus - *Pohlia* Hedw.

Habitat: Grows commonly on peat, sandy banks, on decaying wood, on tree bases.

Leafy gametophyte erect, loosely tufted, branched; leaves erect, narrowly ovate-lanceolate, narrower and decurrent at base; apex apiculate, margin flat entire, very mildly denticulate at the apex; costa percurrent or slightly excurrent; leaf cell at tip linear, leaf base cells broad rectangular; capsule pyriform bottle-shaped, drooping or horizontal, with narrow apophysis; seta arcuate at tip; peristome bryoid type.

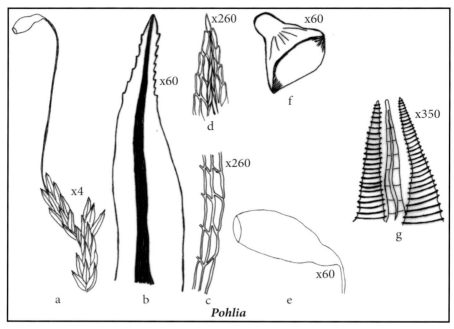

Fig. 4.29. (a) Gametophyte with sporophyte (b) Leaf (c) Alar cells (d) Cells at the tip of leaf (e) Capsule (f) Calyptra (g) Peristome teeth

Order – Bartramiales

Family - Bartramiaceae

Genus – *Bartramia* Hedw.

Habitat: Grows commonly on sandy banks in low lands and rock ledges in mountains.

Leafy gametophyte, tall, in dense tufts, bright green above, brown tomentose below; stem densely covered by leaf; leaves lanceolate – subulate, semisheathing base, divergent-spreading and slightly flexuose or falcate in lamina; margin entire below, sharply toothed in the upper lamina; costa thin, percurrent in the narrow tip; leaf cells thick-walled, short-rectangular at top, longer-narrower at base; capsule rounded to pear-shaped, light brown, seta short, lateral by subsequent growth of stem, flexuose when dry; operculum conical; peristome diplolepidous, deep inserted, outer teeth brownish red, lanceolate; inner teeth alternate, hyaline, fine papillose.

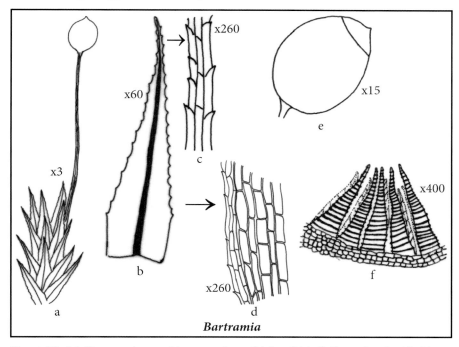

Fig. 4.30. (a) Gametophyte with sporophyte (b) Leaf (c) Cells at the tip (d) Alar cells (e) Capsule with operculum (f) Peristome teeth

Identification key of some commonly occurring genera of mosses (prepared by Late Dr. Prabir Chatterjee)

1. Branches in whorls, leaves with two types of cells – narrow chlorophyllose and wide colourless fibrose   *Sphagnum*
   • Plants not showing branches in whorls, leaves of different areolation   2
2. Leaves distichous or plants flattened (complanate)   3
   • Leaves in three or more rows on stem   5
3. Truly distichous, leaves with characteristic sheathing lamina at the base   *Fissidens*
   • Plants complanate, but not truly distichous   4
4. Leaves single nerved, bordered, plants usually prostrate   *Mnium*
   • Leaves with rudimentary or no nerve, branches pinnate   *Neckera*
5. Leaves with longitudinal lamellae on inner face comparatively stiff, peristome teeth solid (nematodontous)   6
   • Peristome teeth arthrodontous   7

6. Peristome absent, capsule flat, ovoid or ellipsoid, large
    plants   *Lyellia*
    • Capsule smooth, cylindrical, round in t.s., calyptra felt-
      like, peristome teeth 32   *Pogonatum*
    • Capsule ribbed, usually tetragonal or hexagonal   *Polytrichum*
7. Main stem normally erect and short   8
    • Main stem prostrate and long. Peristome double   13
8. Inner basal cells of leaf hyaline and sharply differentiated
    from the normal chlorophyllose lamina cells. On stem
    bark showing gemmae clusters terminal on an elongated
    midrib   *Calymperes*
    • Leaves not hyaline inner base   9
9. Capsule on very short seta, not exserted   *Garckea*
    • Capsule on long exserted seta   10
10. Seta and capsule straight, peristome, if present, single   11
    • Seta usually curved with pendulous capsules, peristome
      when developed double   12
11. Upper leaves in rosette, peristome teeth absent, leaves
    broad, lanceolate or spathulate   *Hyophila*
    • Leaves not in rosette, peristomes spirally twisted or
      very much reduced (*Semibarbula*)   Barbuloid group
12. Upper leaves in rosette, calyptra like a hood (cucullate),
    long-beaked, curved capsule   *Funaria*
    • Leaves narrower, clasping.   *Bryum*
13. Leaf-cells smooth   14
    • Leaf-cells papillose   17
14. Seta long   15
    • Seta short, pinnately branched, leaf-cells increassate,
      vein absent or not well developed   *Neckera*
15. Alar cells distinct   16
    • Alar not distinct, ventral leaves smaller and narrower
      than lateral, nerve very short or missing, capsule
      inclined   *Vesicularia*
16. Capsule erect, alar cells quadrate, seta erect, red brown   *Entodon*
    • Capsule inclined and asymmetric, alar cells incrassate
      (thickened)   *Hypnum*

17. Seta short, long, pendant, pinnately branched epiphytes,
    vein more than ¼ leaf length   *Floribundaria*
    • Seta long, capsule inclined, growing on tree bark, all
      leaves similar, vein absent.   *Taxithelium*

Morphological explanation of some structures

1. Adaxial and abaxial side of leaves of mosses – Leaves of the mosses are attached to the stem like axis transversely. Generally, their apices diverge radially outward from the stem. In acrocarpous mosses, which stand erect, this orientation exposes the adaxial surface to light from above and directs the abaxial surface towards the substrate.

   In pleurocarpous mosses, which are prostrate, the adaxial surface is directed towards the substrate and the abaxial surface is exposed. For this reason, in many keys and descriptions the abaxial surface is defined as the dorsal side of the leaf and the adaxial surface as the ventral side.

2. Antheridiophore – This is a structure, in some complex thalloid liverworts, where the antheridia are concentrated in chambers on the upper surface of a disc like receptacle which is elevated on a stalk formed as a branch system by the prolongation of the main thallus.

3. Archegoniophore – This is a structure, in some complex thalloid liverworts, where the archegonia are initiated and borne dorsally on the receptacle (carpocephallum) – often displaced to the ventral side by the growth of the middle portion of the receptacle – elevated by a stalk formed as a branch system of the thallus.

4. Calyptra – The calyptra is an enlarged archegonial venter that protects the capsule containing the embryonic sporophyte.

5. Elaterophore – In some members of the subclass - Pelliidae a cluster of elaters is attached to the basal end of the cavity of the capsule of sporophyte. Elaterophore can also be apical (Order – Metzgeriales). Here a cluster of elaters is attached to the apex of the capsule. At the point of attachment of these elaters is a small structure which consists of indistinct tissue appearing as an apical thickening of the capsule wall or occasionally as a structure with more or less cellular entity with irregular secondary wall thickening of upto about 35 cells at each valve apex.

6. Incubous arrangement of leaves – In leafy liverworts where the upper border of a lateral leaf (i.e., border facing apex), when viewed from the dorsal surface, overlaps the lower border of the leaf immediately above it on the stem is called incubous arrangement. Ex. *Porella* sp.

7. Involucre – In *Anthoceros* growth of overarching gametophytic tissue occurs forming a protective involucre as the embryo develops. The involucre is ruptured with continued maturation of the sporophyte. The involucre remains as a cylinder that surrounds the base of sporophyte. In the members of the order Sphaerocarpales each sex organ (antheridium or archegonium) is enclosed in a bottle shaped covering with apical opening called involucre. A hood like involucre is formed around the archegonial cluster in *Pellia*. In *Marchantia* one layered two lipped curtain like tissue called involucre is formed on either side of each row of archegonia in archegoniophore.

8. Perianth – In some leafy liverworts a protective covering is formed over the young sporophytes that result from the fusion of two or three leaves around the archegonium is called perianth. Perianth shape and ornamentation vary substantially among leafy liverworts and provide critical taxonomic characters.

9. Perichaetium – In mosses, the archegonia are surrounded and subtended by specialized leaves usually differentiated from the vegetative leaves. The archegonia and their subtending leaves are referred to as perichaetium.

10. Perigonium – In mosses, the antheridia and their subtending leaves are called perigonia.

11. Perigynium – The gametophytic tissue, within the confine of which the sporophyte develops, occurring solely from the female gametophore (mosses) is called perigynium (coelocaule). In *Marchantia* a ring of cells at the base of the calyptra develops to form one layered covering surrounding the growing sporangium called perigynium.

12. Shoot calyptra – When the calyptra and perigynium in combination form the covering of the developing sporophyte then it is called shoot-calyptra.

13. Spore wall of mosses – The spore wall is composed of three distinct layers – an inner intine, a median exine and an outer perine. The intine is made up of polysaccharide. It develops later than the exine and is ultimately greater in thickness. It is made up of fibrillar material. The intine of most mosses is thickened in one particular area where it faces a thinner exine. This area

called the aperture or leptoma is place for emergence of germ tube. The exine is composed of sporopollenin or composed of sporopollenin like compound that confers resistance to desiccation. During early spore development the outermost layer – perine – is deposited upon the exine. The outermost layer perine is contributed by the sporophytic tissue or their break down rather than by the spore themselves. Ornamentation of the outer spore wall comes primarily from the perine, which is formed from deposits of globular materials produced within the spore sac.

14. Succubous arrangement of leaves – In leafy liverworts where the lower border of a leaf, when viewed from the dorsal surface, overlaps the upper border of a leaf immediately below it on the same side. Ex. *Plagiochila* sp.

15. Trigones – Corner wall thickening, made of hemicllulose, between leaf cells of foliose liverworts. They play a role in apoplastic conduction of water.

16. Vitta – Vitta or nerve is a line of highly elongate, thick walled cells found in the leaf of the liverwort genus like *Herbertus* (Jungermanniales).

Chapter 5

Life forms and growth forms of bryophytes

Unlike most of the higher plants, bryophytes are not found as single individuals but in groups of individuals which have characteristic morphological, structural and developmental features. These features help to ascertain, to certain extent, which taxonomical category they belong to – sub-class, order, family, genus or species. The macromorphological features enable the experienced bryologist to identify many genera and often species from quite a distance. The protonema produced from a spore forms one to several buds, each of which grows to become an 'individual'. The individuals are thus at the very outset part of an assemblage. In bryophytes it is not the individual that form the ecological unit but rather the clonal or colonial life-form. Each individual (moss shoot) has a genetically fixed method of ramification, depending on species, genus or family, a particular 'growth form'. Assemblage of individuals and growth-form, modified by external conditions, together provide the characteristics which can be described as the 'life-form'.

Growth-forms

The term 'growth form' is defined by Meusel (1935) as the overall character of a plant that can only be determined by detailed morphological analysis. According to Buchloh (1951) 'the growth forms are fundamentally established in the organization itself'. Growth form is the structural architecture of the individual moss-plant (La Farge England, 1996). It is a purely morphological term.

Growth-forms of mosses (according to Meusel) are as follows :

(i) **Orthotropic mosses:** The stems stand up vertically from the substrate, gametangia positions and sporogonia are acrocarpous. Fertile plants continue their growth by lateral innovation.

  (a) Protonema mosses: The protonema produces short, orthotropic shoots bearing gametangia, which wither after ripening of the sporogonium and are almost always annuals. Ex: *Buxbaumia, Diphyscium.*

  (b) Turf mosses: The sporogonium bearing shoot forms new orthotropic lateral shoots which again bear archegonia and thus sporophytes. Moss turfs of various densities and heights are thus produced. Ex: *Dicranum undulatum.*

(ii) **Plagiotropic mosses:** The shoots lie more or less close to the substrate and can be differentiated into main and lateral shoots. Gametangia are borne on short lateral shoots (pleurocarpous).

(a) Thread moss: There is little difference between main and lateral shoots. This growth form is found in most of the Laskeaceae, *Amblystegium* sp.

(b) Comb mosses: The strong main shoot lying close to or climbing over the substrate bears many simple or branched lateral shoots. Ex: Hypnaceae.

(c) Creeping shoot-mosses: From creeping, rhizome like main shoots branches arise which are upright and stand away from the substrate. They are either single to slightly branched. Ex: Climaciaceae

Growth forms of liverworts

(i) Orthotropic liverworts – *Takakia, Haplomitrium*

(ii) Thread forms – Lophocleaceae, Lejeuneaceae

(iii) Comb forms – *Porella, Frullania*

(iv) Creeping forms – *Bryopteris*

(v) Thalloid forms – Marchantiaceae, Metzgeriaceae

Life-forms

Life-form means, following Warming (1896) who coined this term, the habit of a plant in harmony with its life conditions: this is made up of the components that have already been mentioned, i.e., growth-form – assemblage of individuals and the influence of external factors. Life-forms embodies all selection pressures that are brought to bear upon a species. Mägdefran (1969) described it as "the organization of a plant in correspondence with its life conditions". It is genetically determined. The life-form is so constructed as to minimize evaporative loss while maximizing photosynthetic light capture.

(i) **Annuals**: The gametophyte stops growing once it has produced gametangia and dies after the sporogonium has ripened without having first produced any regenerative shoots. These include pioneer mosses on open mineral soil. Ex: *Buxbaumia, Diphyscium* (Mosses), *Sphaerocarpus, Calobryum, Riccia* (Hepatics).

(ii) **Short turf**: The short shoots, hardly more than 1 cm. high, stand close together and grow on after ripening of the sporogonium by means of regenerative shoots. Ex; *Barbula, Ceratodon* (mosses), *Gymnomitrion* (Hepatics).

(iii) **Tall turf:** The upright shoots, which are not branched or only slightly so, form turf of considerable height. The shoots grow on after gametangia formation. Ex: Dicranaceae.

(iv) **Cushions:** Basal regenerative shoots are produced usually in considerable numbers on the upright shoots. The cushions therefore grow not only upwards but also extend sideways. If they are free-standing they are hemi-spherical in shape. If they are laterally inhibited, they become extraordinarily dense. Ex: *Hypnum*.

(v) **Mats:** Plagiotropic bryophytes, the main and lateral shoots of which lie close to the substrate and are attached to it by rhizoids. Ex: *Taxithelium,* Marchantiaceae, Metzgeriaceae.

(vi) **Wefts:** Plagiotropic bryophytes, the main and lateral shoots of which grow loosely through one another and form a covering that is easy to lift from the substrate, a new layer growing every year over that of the previous year. Wefts form the main constituent of the mossy covering of the forest floors of temperate zones and are able to hold considerable quantities of rainwater by capillary action. Ex: Hypnaceae (Moss), *Lepidozia* (Hepatics).

(vii) **Pendants:** Epiphytic, mostly plagiotropic bryophytes, the main shoots of which hang down from the branches and twigs of trees like beard mosses, whilst the lateral shoots remain short and stand out horizontally. Ex: Meteoriaceae (moss), *Frullania* (Hepatics).

(viii) **Tails:** Bryophytes growing on trees and rocks, mostly shade-loving and of creeping habitat. Their shoots stand out and are slightly branched or unbranched and usually radially leafed. Ex: *Prionodon* (Moss), *Plagiochila* (Hepatics).

(ix) **Fans:** Creeping mosses growing on a vertical base (trees, rocks), the shoots of which branch towards one another in the same plane, project horizontally to obliquely downwards and usually have flattened leaves. Ex: *Thamnobryum* (Moss), *Plagiochila* (Hepatics).

(x) **Dendroids:** Bryophytes growing on the ground, the negatively geotropic shoots of which bear at the top a tuft of large leaves or many lateral shoots. Ex: *Rhodobryum* (Moss), *Pallavicinia* (Hepatics).

(xi) **Steamer:** Long, floating stems in streams and lakes – *Fontinalis sp.*

La Farge-England (1996) defined life-form as the structures and assemblage of individual shoots, branching pattern and direction of growth with modification by its habitat. Growth form applies to the structures of the individual shoot, including direction of growth, combined with length, frequency and position of branches. *Grimmia* cushion is a life-form that has responded to xeric habitat and in a conglomerate of individuals. Its growth form is erect form with variable number of branches positioned along its stem.

Dependency of life-forms on environmental conditions

As in flowering plants, the life-forms of bryophytes are closely connected with two environmental factors: light and water.

Light inhibits elongation of the axes. This causes predominance of short turfs and cushions on brightly lit biotopes. The tails and fans, however, require a dim light. The fans hold their surfaces vertically to the direction of the light, i.e., horizontally to obliquely downwards. This is found in tropical rain forests and cloud forests.

The external, capillary conduction of water is important to those mosses that have sufficient quantities of soil-water available to them. A situation found in bogland and in moist tundra. Here, the life-form of the tall turfs predominates. Because of their crowded shoots, dense foliage and rhizoid weft of these mosses show great efficiency in capillary water conduction.

In the case of mats and wefts and also tails and fans capillary conduction plays insignificant role in the natural habitat but capillary retention does. The full activity, particularly of photosynthesis, is therefore extended considerably beyond the period of precipitation. The cushion life-form is highly adapted for water conservation. Proctor (1980) found that laminar flow patterns over moss cushions were consistent with the measured loss of water from surfaces of varying degree of roughness.

Some moss cushions are so firm that when somebody works on them the foot does not sink into them. 'Moss balls' are a special form of cushion. They are cushions of *Andreaea* and *Grimmia* species which have been levered up from their substrate, probably due to the effects of frost, lie on the soil and rolled hither and thither by the wind and therefore grow radially on all sides.

The striking habitat of pendulous mosses which are limited almost entirely to tropical cloud forests is due to the action of water. The enormous elongation of the main axis is evidently determined by the growing tip which is continually moistened by falling water and thus grows without interruption. Observations on European mosses close to waterfalls (*Plagiomnium, Thuidium*) support this explanation.

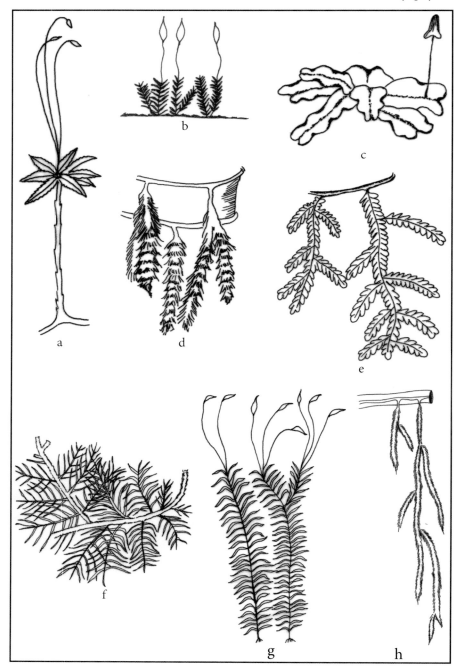

Fig. 5.1. Life form and growth form (a) Dendroid (b) Short turf (c) Mats (d) Tail
(e) Fan (f) Weft (g) Tall turf (h) Pendant

Chapter 6

Moss spore germination

In bryophytes, the dominant phase of life-cycle is haploid and after fertilization a sporophyte with short life span produces spores. Spore germination in most mosses is initiated by the spore swelling through water, followed by exospore rupture, cellular distension accompanied by protrusion of the germ tube and cellular division of the protonema (Schofield, 1985). Two types of germination occur among the bryophytes - (i) in endosporic development the spore cell divides within the cell wall, creating a multicellular structure before protonemal thread emerges from the spore wall. This development occurs while the spore is still within the capsule. This was found in saxicolous mosses (Nehira 1987), (ii) in exosporic development (occurring in most mosses) the development occurs outside the spore wall. Nutrients are important for the development of protonema of mosses (Duckett et al. 2004), but the germination of spores as a physical process can occur in nutrient free medium (for example distilled water) (Olesen and Mogensen 1978). Light may not be essential for the first phase of germination (swelling) but it can be important for the completion of the final phase (cellular division of protonema).

Methodology

(i) Moss plants (gametophytic plant body with sporophyte borne on them) containing mature closed capsules should be taken for experiments. Sporophytic portion (seta with capsule) are separated from the gametophyte and sterilized in 1.5% sodium hypochlorite for 2 minutes and washed in sterile distilled water for 3 – 4 times (Duckett et al. 2004).

(ii) Spores from the capsules are homogenized by the following method – Swollen and brown capsules are transferred to 900 µl sterile distilled water to which is added 100 µl sodium hypochlorite and incubated for 5 minutes. One capsule is taken with the help of a pipette tip and put in a tube containing 200 µl sterile distilled water. Capsule is squashed against the side of the tube to burst it and spores are mixed with the water. 800 µl sterile distilled water is added and the spores are mixed well by pipetting up and down.

(iii) Spores can be put in distilled water or nutrient solution under light or continuous darkness. Three replicates should be used per treatment.

(iv) The Erlenmeyer flasks (100 ml) containing 25 ml distilled water or nutrient solution as detailed below are taken–

$MgSO_4$, $7 H_2O$ – 50 mg L^{-1};

KNO_3 – 120 mg L^{-1};

$Ca(NO_3)_2$, $4 H_2O$ – 1440 mg L^{-1};

KH_2PO_4 – 250 mg L^{-1};

Iron solution – 1 ml/L (Na_2EDTA – 33.2 g/L; NaOH – 3.65 g/L; Fe SO_4, $7 H_2O$ – 25 g/L)] (Dyer, 1979).

Fungicide-Nistatin (100U mL^{-1}) is added.

(v) All the flasks should be covered with gauze and plastic film to avoid evaporation and contamination.

(vi) All the glass wares, nutrient solution and distilled water should be autoclaved at 120°C for 20 minutes.

(vii) The flaks are incubated under 12 hour photoperiod at 25±1°C and light intensity of approximately 20 μmol m^{-2} S^{-1}.

(viii) Spores are considered to have germinated when protrusion of germ tube or protonema with one or more cells are observed.

Germination in Petri plates

(i) Liquid growth medium mixed with 4 gm Agar is poured in Petri dish.

(ii) Growth medium has the following composition –

> **Stock solution 1** - ($MgSO_4$, $7 H_2O$ – 2.5 gm. Add distilled water to make up to 100 ml.).
>
> **Stock solution 2** - (KH_2PO_4) – 2.5 gm. Add distilled water to make up to 50 ml. Adjust pH to 6.5 with 4M KOH.
>
> **Stock solution 3** – KNO_3 10.1 gm; $FeSO_4$, $7 H_2O$ – 0.125 gm. Add distilled water to make up to 100 ml.

Final growth medium to be formed as per the following schedule – 5 ml of stock 1, stock 2 and stock 3. To this added 5 ml of 500 mM Ammonium tartrate and 4 gm of Agar are added. Distilled water to be added to make up to 490 ml. Then the whole mixture is autoclaved. To this added 10 ml. sterile 500 mM $CaCl_2$.

Jian-Cheng et al. (2004) conducted experiments on the characteristics of spore germination in *Lindbergia brachyptera*. The spores were germinated on agar substrate containing modified Knop's solution. Capsules were dipped in 75% ethanol and subsequently cleaned 5 times by distilled water. The spores were mixed with 10 ml distilled water to make the spore fluid. The spore fluid was poured on to the solidified agar substrate with the use of a pipette. Spores were incubated at a temperature of 20±2°C, relative humidity 80%, illumination 24 lux m⁻¹ s⁻¹ and 12 hour day-night cycle.

Composition of modified Knop's solution (pH=7.0)

Reagent.. Concentration (mgL⁻¹)
Ca(NO$_3$)$_2$ 1000
KNO$_3$.. 250
KH$_2$PO$_4$ 250
MgSO$_4$, 7 H$_2$O... 250
ZnSO$_4$, 7 H$_2$O.. 3
FeSO$_4$, 7 H$_2$O 12.5
NaNO$_3$... Subtle
Distilled water.. 1L

Inspection of spore germination

The growing condition of the spores is to be observed everyday. Germinating spores should be randomly selected for observation. The observation was made under the 10x10 microscopic eye pieces. Microphotographs are to be taken at regular interval. Percentage of germination is calculated by the average of 3 observations.

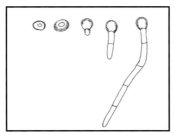

Fig. 6.1. Different stages of spore germination of moss

Chapter 7

Conservation of Bryophytes

The term conservation biology was introduced as the title of a conference held at the University of California, San Diego in 1978. Conservation biology and the concept of biodiversity emerged together to help develop the concept of conservation science and policy. Conservation biology is the scientific study of nature and of earth's diversity with the aim of protecting species, their habitats and ecosystems from excessive rates of extinction and the erosion of biotic interactions. Conservation biology has developed as a discipline of science that involves study of ecology and biodiversity on one side and by active participation of management and environmental measures on the other. Conservation is a process that starts with the identification of threatened habitats/species/genotypes, continues with analysis of threats and finally results in action to ensure the long-term survival of the species.

The bryophytes (liverworts, hornworts, mosses) have great ecological significance in terms of water balance, erosion control, nitrogen fixation, nutrient cycling, providing habitats for other organisms, food chains and animal interaction, facilitation of colonization by vascular plants and form mycorrhizal relationship. Many new chemicals, biochemical pathways and biological processes have been observed in bryophytes and developmental, epigenetic and evolutionary phenomenon have also been documented in the group. In environments like Arctic, the Antarctic, in alpine habitats in mountains above tree line and in bogs, fens and larger peatlands bryophytes form a dominant flora in terms of both biomass and productivity. However, they are inconspicuous elements in many other landscapes because of their small size. Considering their significant role in ecology, the conservation of bryophytes is of great importance. Before I deal with the methods of conservation the ecological significance of bryophytes is discussed in detail.

Nutrient cycling is important in a variety of ecosystems. It is affected by continuing climate change involving changes in temperature, precipitation, in increased atmospheric CO_2 and UV B. Bryophytes capture mineral nutrients by facilitated diffusion which involves ion channels and carrier proteins and depends on the existing gradients of concentration and electric charge across membrane. They frequently accumulate chemicals in much higher concentration than in the ambient environment. This is one important reason for the use of mosses for biomonitoring of air pollution. It has been demonstrated that *Sphagnum* captured

NO_3 during both natural precipitation and experimental treatments more efficiently than rooted vascular plants. It has also been shown that bryophytes are efficient in obtaining nutrients from precipitation as well as from dust and litter before they can be taken up by rooted vascular plants. Bryophytes absorb nutrients and grow during fall and winter while higher plants remain inactive and release nutrients by decomposition in spring and fall. These accumulated nutrients are then used by vascular plants.

Study of the liverwort *Blasia* is of interest. It contains N_2-fixing cyanobacteria *Nostoc* in ventral auricles. *Blasia* as well as *Anthoceros*, which also harbours *Nostoc*, when starved of nitrogen compounds, release a chemical signal that induces formation of hormogonia. They eventually move into the ventral auricles of *Blasia* and after developmental changes generate *Nostoc* filaments with heterocysts. 80% of the N_2 – fixed is leaked to the host *Blasia* in the form of NH_3. *Blasia* is an early successional plant. The plant dies and mosses and vascular plant seedlings take part in the subsequent succession, presumably using the nitrogen leached from the *Nostoc*.

Bryophytes have a high water retention capacity due to their structure and tend to be most abundant in regions with high levels of atmospheric humidity and low rates of evaporation. They can quickly absorb water and release it slowly into the surrounding environment and can contribute to the regulation of water flow. These properties prevent flash floods, erosion and landslides downstream.

Mosses are often the first plants to colonise newly exposed ground, base rocks and other abiotic surfaces. They stabilize soil crust, control erosion and hydric cycling. They play important role as colonisers and soil stabilisers. Bryophytes also help in accumulation of humus.

Bryophyte communities are critical to the survival of a tremendous diversity of organisms, including insects, millipedes and earthworms. The nutrient rich, spore producing capsules are particularly palatable to some insects and molluscs.

Due to lack of a protective cuticle, bryophytes are extremely sensitive to pollutants in the immediate environment. Bryophytes can be used as indicator species as the presence of pollution-sensitive species can help to indicate low levels of air pollution. The mosses are widely used to measure heavy metal air-pollution, especially in large cities and in areas surrounding power stations and metallurgical works. Bryophytes are also suitable for water pollution indicator and also helpful in indicating radioactivity, changes in pH in soil and water, changes in humidity. Some species are strongly associated with calcareous substrates (*Tortella tortusa*) while others will grow only on acid ground (*Racomitrium lanuginosum*). Certain bryophytes have been found to be closely associated with particular mineral or metal deposits such as copper ore. Bryophytes thus can assist in geobotanical prospecting.

Future climate change, particularly in terms of changes in the geographic distribution of precipitation is of course uncertain. Many factor including anthropogenic are important. In some regions it is extremely likely that desiccation will become a problem. Bryophytes may also do better under such circumstances than vascular plants.

Threats to bryophytes

a. Modern agriculture threatens bryophyte diversity through a variety of processes including physical disturbance of the soil by heavy machinery, excessive use of fertilizer resulting in eutrophication of aquatic habitats and excessive herbicide use. Because of the lack of cuticle, bryophyte tissue readily absorbs many different substances from the immediate environment. Bryophyte populations are affected by pollution regardless of whether they are located near to or is far from the area of discharge or in what habitat they are formed. As pollution does not recognize boundaries, bryophytes within protected areas are as susceptible as those outside.

b. Sulphur dioxide is generally the most harmful component of air pollution for terrestrial bryophytes, causing chlorophyll plasmolysis whereas sewage and chemical waste have the greatest effect on aquatic species. Damage may affect sexual reproduction before causing any damage to mature specimens.

c. Bryophytes are efficient accumulators of nutrients, and also absorb heavy metals. Once incorporated these become strongly bound and retained in the tissue. Heavy metals, specifically mercury, lead, copper and cadmium are toxic to nearly all bryophyte species.

d. The collection of bryophytes, whether discriminate or not can pose a threat to the survival of some species. In several countries (U.S.A., India, China) an increase in indiscriminate bryophyte harvesting from natural or semi-natural habitats has been recorded. Large mats of materials are collected, irrespective of species, usually for horticultural use. This threat is not generally monitored by government or international authorities and can result in considerable ecological damage.

e. Collection of specimens by botanists is highly selective on individual species, often varieties. Although this is seldom a real threat to the survival of rare species and modern bryologists tend to recognize that they have a responsibility to conserve there have been cases where collection has led to the extinction of a species. This occurred particularly when botanical collection took place on a much larger scale with little or no regard for the consequences.

f. When bryophytes are directly harvested from natural populations, it is referred to as direct threat. The high water-holding capacity of bryophytes such as *Leucobryum* and *Sphagnum* makes them a useful potting medium, favoured by orchid growers and for wrapping flowers or fruit tree root stocks for transformation. Such kind of bryophytes is directly harvested from nature. If not monitored effectively, such activity may result in considerable ecological damage and decline of bryophyte diversity. Epiphytic moss harvest is most developed in temperate rain forest such as those found along the Pacific coast of Western North America. 15% of the moss species are endemic here and this is an important region for bryophyte conservation. Slow rates of accumulation and the unwanted harvest of non-target species are the reasons to implement proper monitoring of harvest to ensure sustainability and maintain diversity.

g. In addition to direct threats like harvesting, serious threats originate indirectly from the destruction or degradation of their habitats. At the global scale, threat includes factors like – global warming, population and biological invasions. Invasive plants often constitute a threat to biodiversity and have consequently received much attention recently.

At the local scale, deforestation, for both agriculture and timber harvest is one of the most serious areas of concern. Areas like Brazilian Amazon are habitats wich were once continuous (Laurance, 1998). Such areas have become increasingly fragmented. Fragments become isolated from one another by crop land, pasture, pavement or barren land. Fragmentation is one of the main issues in conservation. Epiphytes can be classified into "sun" and "shade" species. Shade epiphytes are characteristic of the under-storey of dense primary forests. They are less desiccation-tolerant than sun epiphytes and generalists that developed a series of putative adaptations, such as papillose cell walls, which enhance the capillary absorption and speed up the process of rehydration. Shade epiphytes are, therefore, highly sensitive to disturbance. The ratio of shade versus light epiphytes is used as a criteria to describe naturalness of forest strands (Gradstein,1992, Acebey et al., 2003).

Methods of conservation

Recently, the importance of *ex situ* conservation and the value of *in vitro* biotechnology have been endorsed in the Convention of Biologhical Diversity (UNEP 1992) and in the subsequent Global strategy for plant conservation (UNEP

2002). In many countries a substantial proportion of the bryoflora worldwide is threatened in the short term (Vaderpoorten and Hallingbäck 2008). Interest in bryophyte conservation has increased significantly in the last two decades.

Some of the recommended measures for conservation after (Hallingbäck and Hodgetts (2000)

(i)   Forests where globally threatened species will occur must be legally protected.

(ii)   Silviculture or partial timber exploitation or both should be prohibited.

(iii)   Felling should be prevented or severely limited in all sites where endangered bryophytes occur.

(iv)   Special attention must be paid to protect important bryophyte habitat on montane sites against inappropriate development.

(v)   Measures should be sought to ameliorate the adverse effects of off-road vehicles and airborne pollution from industrial development and human settlements close to Arctic and Antarctic tundra.

(vi)   All ravines where endangered bryophyte species occur should receive protection against road and bridge construction and the dumping of the waste.

(vii)   Coastal grassland, rocks and thin turf – these areas have many endemic vascular plants and measures for their protection will probably also secure the future of some bryophytes.

(viii) Well developed habitats must be protected from rubbish dumping and urbanization.

(ix)   Inappropriate modern agricultural practices such as excessive use of herbicides and fertilisers should be avoided. Farmers should be offered incentives to use flora-friendly agricultural practices.

(x)   Authorities in area where bryophyte-rich fen is still relatively abundant should be encouraged to recognize the importance of their regions for this internationally threatened habitat.

(xi)   Once cultivated, steppe areas do not regenerate very easily. All remaining natural bryophyte rich steppe, gypsum- or salt rich habitats which harbor threatetned bryophytes must be protected from destruction and exploitation.

(xii)   Bryophytes are especially vulnerable to disturbance. The destruction of seed plant vegetation results in the elimination of the bryophytes that are dependent on that vegetation for shelter. The survival of seed plant vegetation is also linked to the bryophyte vegetation.

Conservation of liverworts in India

About 850 spp. of liverworts have been reported from India. Liverworts play an important role in ecosystem modeling, nutrient cycling, primary production, modification of habitat, pollution detection etc. Bhattacharya (2011) has described in detail the factors responsible for decline of liverwort flora and conservation methods. Factors responsible for decline are – (i) habitat destruction, (2) over exploitation of forest land and utilization of forest products resulting in gradual disappearance of liverworts, (iii) lack of intensive and extensive exploration. Many liverwort habitats are yet to be explored, (iv) global warming – due to global warming, microclimate required for their growth is changing. Temperature and precipitation amount in hilly areas are changing day by day, resulting in loss of microhabitats for liverworts, (v) catastrophic activity – like storm, earthquake, flood, landslides etc. (vi) biological imparities – 80% of the liverwort taxa are dioecious with limited fertilization range. Most of the liverworts are found to reproduce asexually which leads to somatic adaptation of the species instead of genetically. This causes gradual loss in genetic diversity of this group leading to extinction of many taxa. Conservation strategies adopted are as follows – (i) in wild-life sanctuaries and national parks, liverwort habitats are conserved *in situ.* Documentation of liverworts in these areas have been taken, (ii) higher plants serving as habitats of liverworts have been conserved. This also serves the purpose of *in situ* conservation of corticolous and filiicolous liverworts, (iii) *ex situ* conservation involves creation of liverwort garden in the pattern of moss garden, (iv) Conservation is also being done by *in vitro* germination of spores.

Axenic culture of Moss (Sabovljević, Bijelović and Dragićević 2002)

Experiments were conducted on two spp. of *Bryum. In vitro* cultures of *B. capillare* and *B. argenteum* were initiated from apical shoots of gametophytes and spore respectively.

Small shoots of *B. capillare* were separated from soil and rest of the substratum and material was placed in glasses covered with cheese cloth and rinsed with tap water for 30 minutes.

Various dilute solutions of commercial NaOCl (0.5 to 13%) were used for sterilization. Rinsed plant material was kept for 5 minutes in a bleach solution containing a few drops of detergent and rinsed again with sterile distilled water for surface sterilization.

In case of *B. argenteum,* capsules with operculum or operculum and calyptras both were separated from the plant and washed in distilled water. After separating the calyptras, capsules were kept in 9 – 15% solution of commercial bleach for 5 minutes and then rinsed thrice with sterile distilled water. Capsules were opened by sterilized needle and the spores transferred onto a solid medium. Spores in the capsule are already in sterile condition so do not need any further sterilization.

Small shoots of *B. capillare* were isolated aseptically and cultures with tip side up on a solid nutrient medium.

The basal medium contained MS (Murashige and Skoog 1962) mineral salts and vitamins, 100 mg L^{-1} myoinositol, 30 gL^{-1} sucrose, 0.70 % agar and was supplemented with 1.0 mg L^{-1} 2,4 – dichlorophenoxyacetic acid (2, 4 – D) and 2.0 mg L^{-1} Kinetin. The medium pH was adjusted to 5.8 prior to autoclaving at 115°C for 20 minutes. The cultures were grown at 25±2°C under white flurescent light (47 µmol m^{-2} s^{-1} irradiance) and a day/night regime of 16/8 hours. The plants were subcultured in one month interval.

After 10-14 days some changes were observed in the cultures of *B. capillare*, secondary protonemal developed. After a month, the plants gave caulonema. But even after six months plants failed to form buds on protonemal.

The germination of spores were obtained 7 days after their transfer to the medium followed by visible protonema formation within the next 8 days. Development of caulonema and buds as well as regeneration of the whole gametophytes was obtained 3 days after the beginning of spore germination. Small leafy shoots appeared one month after the germination onset.

Axenic culture of liverwort *Marchantia polymorpha* (Vujičić, Cvetić, Sabovljević & Sabovljević 2010).

After collection, the thalli are separated carefully from the mechanical impurity placed in glasses, covered with cheese cloth and rinsed with tap water for 30 minutes. Gemmae and thallus parts were then disinfected for 5 minutes with a 3 – 15 % solution of sodium hypochlorite. Finally, they were rinsed 3 times in sterile distilled water.

As a basal medium for establishment of *in vitro* culture Murashige and Skoog (1962) (MS) medium containing MS mineral salts and vitamins, 100 mg/l inositol, 0.7 (w/v) agar and 3% sucrose and BCD medium were used.

Once the culture was established, *in vitro* developed plantlets were used for further developmental experiments. Light was supplied by cool-white fluorescent tubes at a photon fluency rate of 47 µmol/m^{2}s. Cultures were subculturd for a period of 4 – 6 weeks.

Cryopreservation

Cryopreservation is often used for the long-term storage of plant germplasm where conventional methods are inappropriate. Simple encapsulation-dehydration protocols were developed for the cryopreservation of bryophytes at the Royal Botanic garden, Kew, as part of a *ex situ* project for the conservation threatened species of UK. These methods were tested on 22 species with a broad range of ecological requirements and found to be highly successful across the board. According to Rowntree and Ramsay (2009) methods applied are as follows –

(i)   Gametophytic material from 21 moss species was obtained from established tissue cultures and prepared for long-term storage in liquid nitrogen (cryopreservation) using a dehydration-encapsulation technique and pre-treatment with 10 μM abscissic acid and 5% (w/v) sucrose (Burch and Wilkinson 2002).

(ii) Small amounts of protonemal material were encapsulated in alginate strips, while single shoot tips from leafy gametophores were in alginate beads.

(iii) Approximately ten strips or beads were prepared together and placed into single cryovials as replicates. Once prepared, cryovials containing bryophyte material were rapidly immersed in liquid nitrogen and stored for 1 week at -196°C.

(iv) The alginate beads or strips from a single replicate were then removed from the liquid nitrogen warmed for 2 months in a water bath at 40°C and transferred into 9 cm Petri-dishes containing growth medium suitable for the individual species (1/4, ½ MS or Knop's minimal medium; pH 5.8, solidified with Gelrite, (Rowntree 2006).

(v)   Regeneration and growth of the bryophytes was monitored thereafter on a weekly basis for a maximum of 10 weeks.

Regeneration rates were extremely high across all of the species tested. It appears that bryophytes and in particular mosses are particularly well suited for storage via cryopreservation. This is due, in part, to totipotency of bryophytes and that many are able to survive extreme, particularly cold and drying environment.

Plate-I : (a) & (b) Thallus of *Lunularia* sp. showing crescent shaped gemma cup and wavy margin **(c)** *Marchantia* thallus forming mat with gemma cup **(d)** *Marchantia* sp. showing archegoniophore (⟹) and antheridiophore (⟹)

Plate-II : (a) Archegoniophore of *Marchantia* sp. showing position of sporophyte (➡) **(b)** Archegoniophore of *Asterella* sp. showing perianth forming a cone (➡) **(c)** Thallus of *Plagiochasma* showing young archegoniophore (➡)

a

b   c

Plate-III : (a) & (b) *Plagiochasma* thallus with mature stalked archegoniophore arising from behind the apical point **(c)** Thallus of *Conocephalum* showing air pores

79

Plate-IV : (a) & (b) Thallus of *Riccia* sp. showing rosette formation **(c)** *Cyathodium* sp. **(d)** *Dumatiera* sp. thallus showing archegoniophore

a

b

Plate-V : (a) Mixed population of *Conocephalum conicum* and *Marchantia linearis* (archegoniophore) **(b)** Mixed population of *Asterella* sp. and *Marchantia* sp. forming mat

a

b

c

Plate-VI : Foliose liverworts – **(a)** *Porella* sp. **(b)** *Plagiochila* sp. **(c)** *Lejeunea* sp.

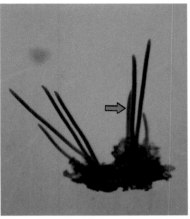

a   b

c

Plate-VII : (a) & (b) *Anthoceros* sp. – thallus with sporophyte **(c)** *Notothylus* sp.
– thallus with sporophyte

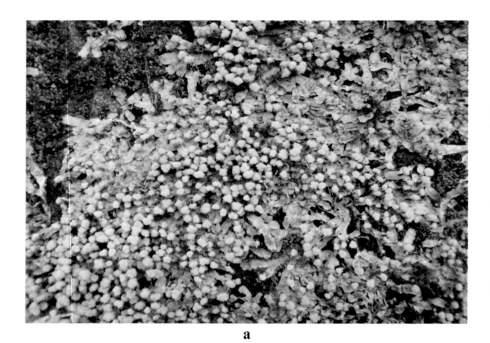

a

b

Plate-VIII : (a) Mat formation by *Asterella* sp. **(b)** Dense growth of *Anthoceros* sp. on soil

84

a

b   c

Plate-IX : (a) *Sphagnum* sp. with capsule **(b)** *Atrichum* sp. **(c)** *Pogonatum* sp.

85

a

b

Plate-X : (a) *Trematadon* sp. **(b)** *Pohlia* sp.

86

Plate-XI : (a&c) *Bryum* sp. **(b)** *Hyophila* sp. **(d)** *Bartrania* sp. **(e)** *Anomoloryum* sp.

a

b

c

Plate-XII : (a) *Semibarbula* sp. **(b)** *Mnium* sp. **(c)** *Physcomitrium* sp. with cap like calyptra with narrow tip

88

a

b

Plate-XIII : **(a)** *Bryum* sp. **(b)** *Anomobryum* sp.

89

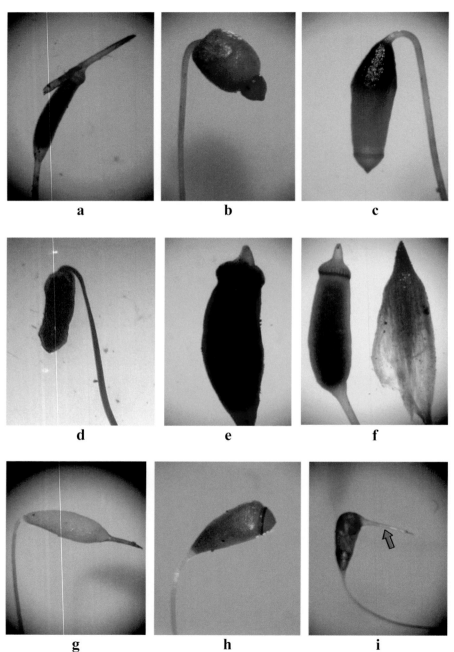

a   b   c

d   e   f

g   h   i

Plate-XIV : Types of moss capsule – **(a)** *Atrichum* sp. with cuculate calyptra **(b), (c), (d)** pendulous capsules of *Bryum* spp. showing different shapes of capsule **(e) & (f)** capsule of *Pogonatum* and Calyptra **(g)** *Mnium* sp. **(h) & (i)** asymmetric capsule of *Funaria* with calyptra

90

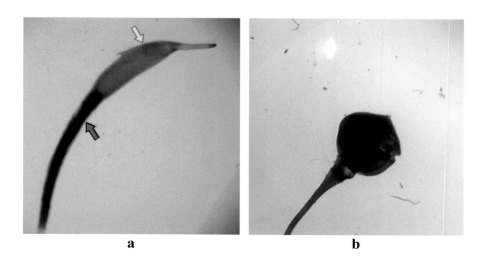

a   b

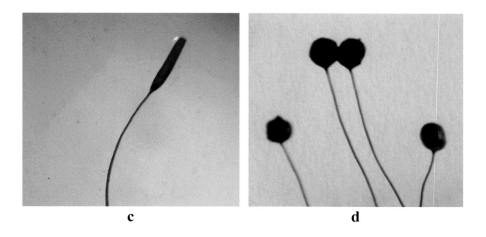

c   d

Plate-XV : Types of capsule – **(a)** *Trematadon* sp. – narrow cylindrical apophysis (➡) and cuculate calyptra(➡) **(b)** globose capsule of *Physcomitricum,* **(c)** *Hyophila* capsule cylindrical, erect **(d)** *Bartramia* sp. – rounded pear shaped capsule

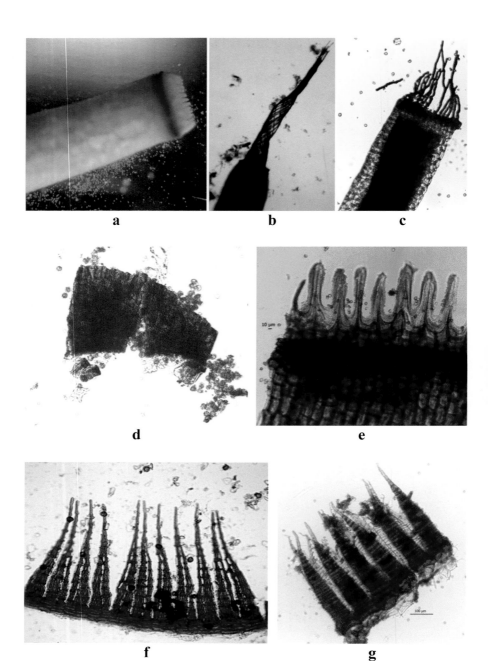

Plate-XVI : Types of peristome teeth – **(a)** Capsule with opened nematodontous peristome teeth **(b)** Capsule with spirally twisted with one-more turns **(c)** Filamentous peristome teeth not twisted **(d)** Undeveloped peristome **(e)** Nematodontous peristome teeth **(f)** Arthrodontous, haplolepidous peristome teeth **(g)** Arthrodontous diplolepidous peristome teeth

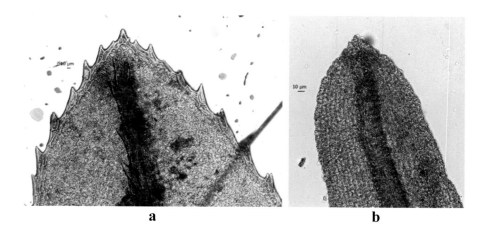

a   b

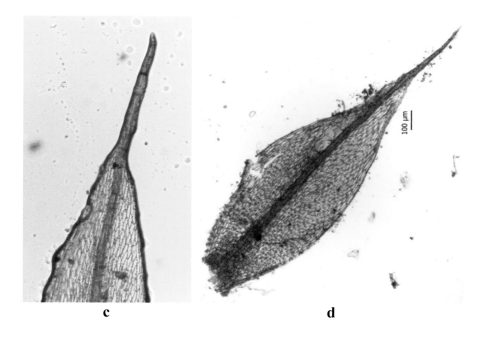

c   d

Plate-XVII : Different aspects of moss leaf morphology – **(a)** Apical portion showing serrated margin, costa percurrent **(b)** Leaf tip broadly pointed, margin smooth below, denticulate at apex, costa percurrent **(c) & (d)** Leaf with excurrent costa

93

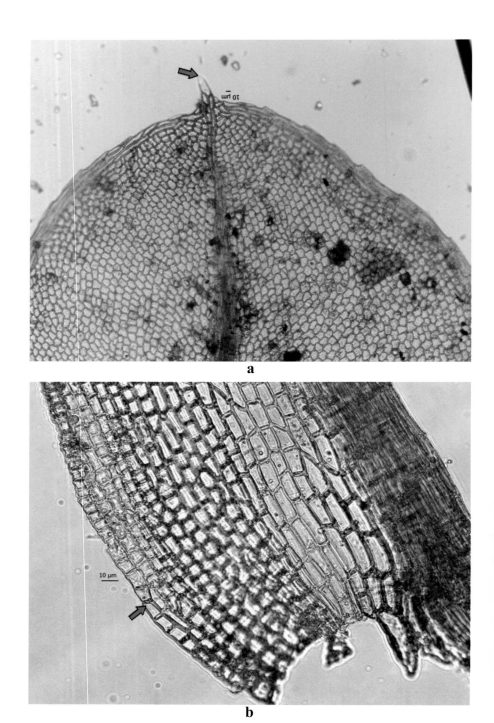

Plate-XVIII : (a) Notched leaf tip of *Mnium* **(b)** Alar cells of moss leaf

94

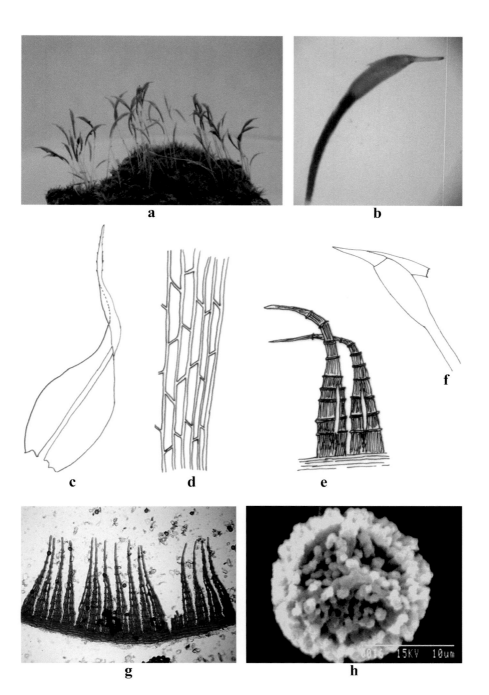

Plate-XIX : A model plate to represent various features of a moss plant – **(a)** Habit, **(b) & (f)** Capsule with calyptra **(c)** Leaf, **(d)** Alar cells **(e) & (g)** Peristome teeth **(h)** Spore ornamentation by SEM

95

a

b

Plate-XX : (a) *Funaria hygrometrica* population on rock **(b)** Moss population on brick wall

96

a

b

Plate-XXI : (a) Moss plant on rock **(b)** Epiphytic moss

97

a

b

Plate-XXII : (a) *Physcomitrium* on garden soil **(b)** New gametophore developing on rocky soil within drying population

98

Glossary

A

abaxial: of the side or surface of an organ, facing away from the axis. cf. **adaxial**.

acaulescent: lacking a stem.

acrocarpous: with the gametophyte producing the sporophyte at the end of the stem or main branch. Most acrocarpous mosses grow erect in tufts, and they are not or only sparsely branched. cf. **pleurocarpous**.

Acrogynous: having the stem terminated by archegonium.

acumen: a slender, tapering point. adj. **acuminate**.

acute: terminating in a distinct but not protracted point, the converging edges separated by an angle of 45–90°.

adaxial: of the side or surface of an organ, facing towards the axis. cf. **abaxial**.

adherent: showing union of parts usually separate.

adnate: united with another organ.

adventitious: produced abnormally.

alar cells: specialised cells at the basal angles of a leaf, often distinctive in their size, shape, colour or ornamentation.

alveolate: with cavities on the surface (spores).

amentula: applied to the special antheridia bearing branches of *Sphagnum*.

amphigastria (sing. **amphigastrium**): leaves that grow in a row on the lower side of a stem and which are usually smaller and have a different shape to other leaves.

amphithecium: the outer embryonic tissue of an embryonic capsule surrounding the central **endothecium**; gives rise to all tissues from the epidermis to the outer spore sac; also produces the spore sac in *Sphagnum*.

amplexicaul: clasping a stem.

anacrogynous: stem not terminated by archegonium.

analogous: structures or organs with similar functions that do not have a common phylogenetic origin; e.g. stomata and air pores. cf. **homologous**.

androecium (pl. **androecia**): the 'male gametoecium' consisting of antheridia, paraphyses and surrounding bracts. See also **perigonium**.

androgynous: with antheridia and archegonia in the same cluster of leaves.

anisomorphic: describing related structures that exhibit more than one distinct type of size or shape.

anisophyllous: having dissimilar stem and branch leaves. e.g. in *Sphagnum* and *Thuidium*; or bearing two distinct types of leaves on the same stem, e.g. in *Hypopterygium* and *Racopilum*.

annual: a plant that completes its life history within one year.

annular: shaped like a ring; leaves or branches arranged in a circle, e.g. *Philonotis*.

annulus: one or more rings of enlarged, specialised cells between the mouth of the capsule and operculum, aiding in dehiscence.

antheridium (pl. **antheridia**): the male gametangium; a multicellular stalked, structure with a jacket of sterile cells and producing large numbers of antherozoids (male gametes); globose to broadly cylindrical in shape.

antherozoid: a motile male gamete; in mosses propelled by two flagellae.

antical: upper surface (stem or leaf).

anticlinal: oriented perpendicular (rather than parallel) to the surface. cf. **periclinal**.

apical cell: a single cell at the apex of a shoot, leaf or other organ that divides repeatedly to produce new leaves, stems or other organs.

apiculus: a short, abrupt point; adj. **apiculate**.

apophysis (pl. **apophyses**): a differentiated sterile neck at base of the capsule, between the seta and urn; sometimes swollen or expanded (= **hypophysis**).

apoplastic movement: the movement of water into a cell via the protoplast, being controlled by osmosis.

appendiculate: having short, thin transverse projections, e.g. on the endostomial cilia of the peristome (see **trabeculae**).

appressed: closely applied, as for leaves lying closely or flat against the stem.

archegonium (pl. **archegonia**): the female gametangium; a multicellular, flask-shaped structure consisting of a stalk, a swollen base (venter) containing the egg and a neck through which the antherozoid swims to fertilise the egg.

arcuate: curved like a bow.

areolate: with small angular areas forming a network; the cellular pattern of the moss leaf is often termed areolation.

arista: the hard awn or bristle at the tip of a leaf, usually formed by an excurrent costa. adj. **aristate**.

arthrodontous: *of a peristome*, of triangular or linear teeth/segments consisting of differentially thickened wall-pairs. The teeth/segments are formed of part cells, in contrast to a**nematodontous** peristome in which they are formed of whole cells.

Articulate: peristome teeth marked by cross bars.

Astomous: capsule without a mouth, used of capsules which has no regular dehiscent lid.

attenuate: tapering gradually.

auricle: a small bulge or ear-like lobe at the basal margin of a leaf, e.g. in *Papillaria* and *Calyptothecium*; adj. **auriculate**.

autoicous: with male and female gametoecia on separate stems or separate branches of the same plant (**monoicous**). cf. **synoicous, paroicous, dioicous**.

awn: an arista or hairpoint, usually hyaline and formed of a projecting costa.

axil: the angle between the stem and any organ originating from it, e.g. a leaf or branch. adj. **axillary**.

axillary hair: a hair originating in a leaf axil, usually inconspicuous and often concealed by the leaf base.

axis: the main stem; the conceptual line around which leaves, branches and other organs develop.

B

basal membrane: a delicate or robust membrane at the base of the endostome, often bearing segments and cilia (= **basement membrane**).

basement membrane: see **basal membrane**.

bifurcate: forked into ±equal parts.

bilabiate: two lipped.

bipartite: divided wanly to the base into two portions.

biseriate: in two rows.

bistratose: consisting of two cell layers, e.g. a leaf lamina two cells thick.

border: *of leaves or the edges of peristome teeth*, a margin that is differentiated in shape, size, colour or thickness. adj. **bordered**.

bract: one of the specialised leaves surrounding and protecting sex organs.

bulbil: a small deciduous, bulb-shaped, axillary, vegetative propagule or rhizoidal gemma; often with rudimentary leaves.

bulbiform: bulb-shaped.

C

caducous: falling readily or early.

caespitose: tufted, growing in dense cushions or turfs.

calcicolous: a plant that grows best in habitats or on substrata with high levels of calcium.

calcifugous: a plant that cannot toletate habitats or substrata with high levels of calcium.

calyptra (pl. **calyptrae**): a membranous or hairy hood or covering that protects the maturing sporophyte; derived largely from the archegonial venter.

campanulate: shaped like a bell; a campanulate-cucullate calyptra is split on one side only, whereas a campanulate-mitrate calyptra is undivided or equally lobed at the base.

canaliculate: channeled.

capillary: hair like.

capitate: Head like.

capitulum (pl. **capitula**): a head-like mass of crowded branches at the apex of the stem, e.g. in *Sphagnum*.

capsule: the terminal, spore-producing part of a moss sporophyte.

carinate: folded along the middle, like the keel of a boat; V-shaped in cross-section.

carpocephalum: female receptacle.

caulonema: a secondary, bud-generating part of the filamentous moss **protonema**, typically reddish brown with few chloroplasts and consisting of long cells with oblique end walls.

central strand: the column of elongated cells, sometimes with thicker walls, in the centre of a stem.

cernuous: nodding or drooping.

channelled: *of a leaf*, hollowed out like a gutter and semicircular in cross-section.

chloronema: the filamentous part of the protonema that contains chloroplasts.

ciliate: fringed with hair.

ciliolate: fringed with very small cilia.

circinate: coiled up into a ring completely or partially.

cladocarpous: having a fruit terminating a lateral shoot.

clavate: club-shaped.

cleistocarpous: *of a capsule*, lacking an operculum and, therefore, opening irregularly.

clone: population of genetically identical plants produced vegetatively from a single propagule or spore.

cochleariform: round and deeply concave, like the bowl of a spoon.

collenchymatous: cells with walls that are thickened at the corners, e.g. exothecial cells or cortical cells of stems seen in cross section.

columella: the sterile, central tissues of a capsule.

commissure: the line of junction of the antical and postical lobes of a leaf.

comose: stems tips with leaves that are larger and crowded into tufts (**comae**), e.g. in *Bryum* and *Campylopus*.

complanate: a leafy shoot that is more-or-less flattened into one plane.

complicate: folded leaf.

concolorous: having the same colour.

conduplicate: folded lengthwise along the middle.

confluent: running into one another.

connate: united.

connivent: converging.

conspecific: belonging to the same species.

constricted: abruptly narrowed.

Contiguous: in contact.

contracted: abruptly narrowed or shortened.

convolute: of leaves or bracts, rolled together to form a sheath, e.g. the perichaetial leaves of Holomitrium.

cordate: heart-shaped, as in leaves attached at the broad end.

coriaceous: leathery in texture.

cortex: the outermost layer or layers of cells in a stem, often differentiated from the central cylinder. adj. **corticate, cortical.**

corticolous: growing on bark.

costa (pl. **costae**): the thickened midrib or nerve of a leaf; when present, can be single or double. adj. **costate.**

crenate: *of a leaf margin,* having rounded teeth.

crenulate: *of a leaf margin,* having minute, rounded teeth formed from bulging cell walls.

crisped (or **crispate**): wavy; often used loosely to include curled, twisted and contorted.

cristate: having a crest-like ridge.

cruciate: like a cross.

cucullate: hooded or in the shape of a hood; applied to leaves that are concave at the tips and to calyptrae that are conical and split up one side.

cuneate: wedge shaped.

cushion: a more-or-less hemispherical or rounded moss colony, with stems generally erect and tightly clustered but radiating somewhat to form a tuft.

cuspidate: ending in a stout, rigid point, like a tooth.

cuticle: a non-cellular coating on the outer surface of cells in contact with the environment, often variously roughened or ornamented.

cygneous: curved downards in the upper part like the neck of a swan, e.g. setae of *Campylopus.*

cylinder: the central strand in stem. adj. **cylindrical.**

cymbiform: concave and boat-shaped.

D

deciduous: falling off, lost at maturity, e.g. the operculum.

decumbent: tips ascending from a prostrate base.

decurrent: applied to the margins of leaves which extend down the stem, as ridges or narrow wings, below the insertion of the leaf.

decurved: curved downward.

deflexed: bent downward.

dehiscent: of capsules, splitting open by means of an annulus, operculum or valves (as opposed to **indehiscent**).

deltoid: triangular.

dendroid: with the habit of a tree, branching from a main stem.

dentate: with teeth directed outward.

denticulate: with fine teeth.

depauperate: stunted or poorly developed.

depressed: flattened, as viewed from above.

descending: directed gradually downward.

diaspore: an agent of dispersal; any structure that becomes detached from the parent plant and gives rise to a new individual.

dichotomous: with two equal forks or branches.

dimorphic: of two distinct forms, e.g. leaves, male and female plants.

dioicous: with archegonia and antheridia borne on separate plants.

diplolepidous: a form of arthrodontous peristome having two concentric rings of teeth, with the outer ring (**exostome**) derived from thickening of the contiguous walls of the outer and primary peristomial layers and the inner ring (**endostome**) derived from the thickening of the contiguous walls of the primary and inner peristomial layers. The exostome is generally more heavily thickened than the endostome. One or both rings may be absent or reduced (cf. **haplolepidous**).

distal: away from the base or point of attachment; the converse of **proximal**.

distant: widely spaced, e.g. leaves with space between adjacent leaves.

distichous: leaves alternating in two opposite rows on a stem, as in *Fissidens*.

divergent: spreading in opposite directions.

dorsal: belonging to or on the back that is a face of leaf remote from the stem, upper surface of thallus.

dorsal lamina: part of the leaf blade opposite the sheathing base, at the back of the costa and below the apical lamina in *Fissidens*.

dorsiventral: flattened with distinct upper and lower surfaces.

E

echinate: bearing spiny projections (spores).

ecostate: lacking a costa.

ectohydric: having water transport essentially external by surface flow, including capillary motion between leaves or through surface papillae. cf. **endohydric**.

efibrillose: without fibrils.

elaters: sterile filaments or cells mixed with spores in capsule.

ellipsoidal: a solid with an elliptical profile.

elliptical: having the shape of an ellipse, oblong but convex at the sides and ends.

emarginate: broad at the apex with a shallow notch.

embryo: the developing sporophyte phase generated from a zygote.

emergent: partly exposed, as a capsule only partly protruding from among the perichaetial leaves. cf. **exserted, immersed**.

endemic: restricted to one country or one floristic region.

endogenous: arising from deep seated tissue.

endohydric: having water transport essentially internal. cf. **ectohydric**.

endostome: the inner ring of a diplolepidous peristome, formed from contiguous periclinal wall-pairs of the primary and and inner peristomial layers; typically a weak membranous structure consisting of a basal membrane bearing **segments** and **cilia**; homologous with the single peristome of haplolepidous mosses.

entire: with a smooth outline, not toothed or lobed.

ephemeral: short-lived.

epidermis: the outer layer of cells at the surface of an organ, e.g. **exothecium**.

epiphragm: a circular membrane, positioned horizontally over the capsule mouth of some mosses, attached to the tips of the peristome teeth and partially closing the mouth of an inoperculate capsule, e.g. *Funaria* , *Polytrichum*.

epiphyllous: a plant that grows on the living leaves of another plant.

epiphyte: a plant that grows on the surface of another plant.

equidistant: regularly separated or spaced.

erect: of leaves, almost or quite parallel to the stem, but not appressed; *of branches or stems*, in a ±vertical position with respect to stem or substratum; *of capsules*, upright.

erecto-patent: spreading at an angle of less than 45°. cf. **spreading** or **patent**.

exannulate: lacking an annulus.

excavate: hollowed out.

excurrent: *of a costa*, extending beyond the leaf apex.

exine: the outermost wall layer of the spore.

exogenous: arising from superficial tissue.

exostome: the outer circle of the diplolepidous peristome, consisting of teeth formed from contiguous periclinal wall-pairs of the outer and primary peristomial layers; absent or rudimentary in the haplolepidous peristome.

exothecium: the epidermis or superficial layer of cells (**exothecial** cells) of the capsule wall.

exserted: exposed, as in a capsule protruding beyond the perichaetial leaves. cf. **emergent**.

F

failing: *of a costa*, terminating below the leaf apex.

falcate: curved like a sickle.

falcato-secund: strongly curved and turned to one side.

fascicle: a group, bunch or tuft of branches, e.g. in *Sphagnum*. adj. **fasciculate**.

fastigiate: with branches erect and of similar length.

fenestrate: pierced with broad openings resembling windows.

fibril: a fine, fibre-like wall thickening. adj. **fibrillose**.

filamentous: thread-like.

filiform: thread-like.

fimbriate: fringed, generally eroded with radiating cell walls of partly eroded marginal cells. cf. **laciniate**.

flabellate: shaped like a fan.

flaccid: soft and limp.

flagelliform: whip-like; a branch with a gradual attenuation from ordinary leaves at the branch base to vestigial-branched tip. cf. **stoloniferous**.

flagellum (pl. **flagella**): a slender, tapering branch; also the organs of locomotion in an antherozoid; adj. **flagellate**.

flexuose: slightly bent, wavy or twisted.

foliose: leafy or leaflike; covered with leaves.

foot: the basal organ of attachment and absorbtion for the bryophyte sporophyte, embedded in the gametophyte.

fringed: with a short-ciliate margin or edge.

frond: the branched or leafy part of an erect stem, including branches of a dendroid moss. adj. **frondose**.

fugaceous: quickly or readily falling or vanishing.

funiculate: rope-like, e.g. of leaf arrangement in some *Macromitrium* spp.

frurcate: forked.

fuscous: dull brown.

fusiform: narrow and tapering at each end, spindle-shaped.

G

galeate: shape like a helmet.

gametangium: an **antheridium** or **archegonium**; a structure forming gametes (**ovum**, **spermatozoid**).

gamete: a haploid reproductive cell, e.g. **spermatozoid**, **ovum**.

gametophore: loosely used for the leafy moss gametophyte plant developed from a protonema.

gametophyte: the haploid, sexual generation; in bryophytes the free-living, dominant generation.

geminate: in pairs.

gemma: uni- or multi-cellular, globose, clavate, filiform, cylindrical or discoid structures, borne on the aerial part of the plant and functioning in vegetative reproduction.

gemmiferous: bearing gemmae.

geniculate: bent abruptly, as at the knee.

gibbose: swollen or bulging at one side.

glabrous: smooth, not papillose, rough or hairy.

glaucous: bluish green in colour or with a greyish or whitish bloom.

granulose: minutely grainy, roughened with minute blunt projections.

gregarious: growing close together in loose tufts or mats.

guard cells: specialised photosynthetic cells bordering the stoma on the capsule wall.

guide cells: large, rather thin-walled cells in the centre of the costa, usually best seen in transverse section.

gymnostomous: without a peristome, so that the mouth of the urn is naked.

H

hairpoint: the hair-like and often colourless leaf tip, formed from an excurrent costa or a tapering of the leaf lamina.

haplolepidous: a form of arthrodontous peristome having only one circle of teeth derived from thickening of the contiguous walls of the primary and inner peristomial layers.

heteroicous: having several forms of gametoecia on the same plant; also called polygamous, polyoicous.

heterolepidous: a form of arthrodontous peristome thought by some to be intermediate between **haplolepidous** and **diplolepidous**, e.g. in *Encalypta*.

heteromallous: pointing in various directions. cf. **homomallous**.

heteromorphic: having two or more different shapes or phases.

hoary: greyish or whitish, appearing frosted from numerous massed hairpoints.

homologous: structures or organs with a common phylogenetic or developmental origin, but not necessarily similar in appearance and/or function. cf. **analogous**.

homomallous: pointing in the same direction. cf. **heteromallous**.

hyaline: colourless and transparent; commonly used with reference to cells that lack chloroplasts.

hyalocyst: a large, hyaline, water-storage cell in Sphagnopsida.

hyalodermis: in *Sphagnum*, an cortex of large, empty, colourless cells. adj. **hyalodermal**.

hydroid: a water-conducting cell in the central strand and/or costa of some mosses, e.g. Polytrichales.

hydrome: a sheath of hydroid cells in the central strand and/or costa of some mosses, e.g. Polytrichales.

hypnoid: having a complete peristome; occasionally used to refer to a moss with a pleurocarpous habit.

hypogynous: inserted below the archegonium.

hypophysis: see **apophysis**.

I

imbricate: closely appressed and overlapping.

immarginate: of a leaf, lacking a border.

immersed: submerged below the surface; immersed capsules occur below the tips of the perichaetial leaves; immersed stomata have guard cells that are sunken below the surrounding exothecial cells.

incised: cut sharply.

inclined: applied to a capsule that is tilted between the vertical and horizontal.

incrassate: thickened, or with thick walls.

incubous: the oblique insertion of distichous leaves show that the lower overlap the other on the same side of the stem on the dorsal surface.

incumbent: lying against or leaning on something.

incurved: curved upward and inward, the opposite of **recurved**; applied to leaf margins and tips.

indehiscent: *of capsules*, lacking a distinct opening mechanism; spores shed by irregular rupture or breakdown of capsule wall, e.g. in *Archidium*.

inflated: swollen, puffed up.

inflexed: bent upward (adaxially) and inward, the opposite of **reflexed**; applied to leaves, leaf margins and peristome teeth.

intra foliar: below the leaves.

initial: an undifferentiated, meristematic cell that divides to produce discrete organs, e.g. rhizoid initial, stem initial or leaf initial.

innovation: a new shoot; in acrocarpous mosses a subfloral branch formed after differentiation of the sex organs.

inoperculate: lacking an operculum.

insertion: a line or point of attachment of a leaf, branch or peristome etc.

Interfoliar: between the leaves.

Intercalary: growth not apical but between the apex and the base.

intine: the innermost wall of the spore.

intramarginal: submarginal; structures close to or associated with but not strictly on the margin.

intricate: tangled, interwoven.

Involucre: a tubular structure serving to protect the archegonia and calyptra.

involute: strongly rolled upward (adaxially) and tightly inward, opposite of **revolute**; applied to leaf margins.

isodiametric: about as long as broad and having the same dimensions in all directions; applied to square, rounded or hexagonal cells.

isomorphic: *of spores*, ±uniform in size.

isophyllous: having similar stem and branch leaves. cf. **anisophyllous**.

J

julaceous: smoothly cylindrical; applied to shoots with crowded, imbricate leaves.

juxtacostal: the part of a leaf lamina adjacent to the costa.

K

Keel: a ridge like the keel of a boat.

L

lacerate: deeply and irregularly cut or torn.

laciniate: dissected into fine, deep, often irregular divisions (**laciniae**); fringed with cilia.

lacunose: when the surface is covered with depressions, perforated with holes.

lamella (pl. **lamellae**): a longitudinal chlorophyllose ridge or plate on the leaf blade of some mosses (e.g. Polytrichaceae); adj. **lamellate**; the plates of the secondary wall deposition occurring between trabeculae on the dorsal and ventral surfaces of an arthrodontous peristome.

lamina (pl. **laminae**): the blade of a leaf excluding the costa and leaf margin or border.

laminal cell: any cell of the lamina.

lanceolate: shaped like the blade of a spear, narrow and tapered from near the broader base.

lax: soft or loose, commonly referring to a tissue of large, thin-walled cells as well as the spacing of leaves.

lenticular: shaped like a double-convex lens.

leptoid: a conducting cell similar in form and function to a sieve tube in vascular plants; found in the central strand and setae of Polytrichales and in the setae of many mosses.

leptome: a tissue, similar to the phloem of vascular plants, consisting of leptoids and parenchymatous cells.

leucocyst : a large, empty hyaline cell in the leaves of *Sphagnopsida* and *Leucobryum* (= **hyalocyst**).

ligulate: strap-shaped, with parallel sides and an abruptly tapered apex.

limb: the upper part of the leaf, the lower part being the **base**.

limbidium: a leaf border or differentiated margin in e.g. *Fissidens*.

linear: very narrow and elongate, with the sides nearly parallel; narrower than **ligulate**.

lingulate: tongue-shaped; broad with the sides ±parallel.

lumen (pl. **lumina**): the cavity of a cell.

M

m-**chromosome:** the smallest chromosome, less than half the length of other members of the chromosome complement; common in bryophytes.

macronema (pl. **macronemata**): a large, branched rhizoid produced around branch primordia and at the base of buds.

mammilla (pl. **mammillae**): a bulge on the surface of cell with a nipple-like tip. adj. **mammillose**.

marsupium: the fruiting receptacle of certain liverworts.

mat: a densely interwoven, horizontal growth form.

median: central, in the middle; **median leaf cells** are those in the upper middle of the leaf or, in leaves with a costa, those located between the margin and costa about two-thirds of the way up the leaf.

meristem: a permanent or temporary zone of actively dividing undifferentiated cells which by, mitotic division, give rise to tissues and organs.

meiosis: the process of nuclear division by which a diploid nucleus yields 4 haploid nuclei; in mosses meiosis takes place in the spore sac of the capsule to produce 4 haploid spores.

mitrate: *of a calyptra*, conical and undivided or regularly lobed at the base.

monoicous: bisexual, having antheridia and archegonia on the same plant; includes **autoicous**, **synoicous** and **paroicous**.

monopodial: with the main stem having unlimited growth, and giving rise to numerous, secondary, lateral shoots or stems.

mucronate: abruptly pointed by a short spinous process.

multifid: divided into many lobes.

muricate: rough with spinous processes.

N

naked: lacking covering structures or ornamentation; e.g. without hairs or papillae, referring to smooth, glabrous calyptra.

neck: the sterile basal part of moss capsule; also the cylindrical upper part of an archegonium.

nematodontous: *of a peristome*, consisting of whole dead cells with ±evenly thickened walls, e.g. as in Polytrichaceae. cf. **arthrodontous**.

nodose: knotted, with small knob-like thickenings; e.g. endostomial cilia in Bryaceae. dim. **nodulose**.

nutant: nodding or drooping.

O

ob: a prefix indicating inversion, as in **obovate**.

obconical: inversely conical.

obcordate: inversely heart shaped.

obcuneate: inversely cuneate.

oblate: wider than long.

oblong: rectangular but, when applied to leaves, usually rounded at the corners.

obovate: with the profile of an egg, the broad end distal.

obtuse: broadly pointed, at an angle of greater than 90°; sometimes used loosely to indicate blunt.

ochraceous: brownish yellow.

operculum (pl **opercula**): the lid covering the mouth of most moss capsules, becoming detached at maturity; usually separated from the mouth by an annulus. adj. **operculate**.

ostiole: the tubular neck of the cavity containing antheridia.

ovate: with the profile of an egg, the base broader than the apex and about twice as long as wide.

P

palmate: having radiating branches originating from a single point.

panduriform: shaped like the body of a violin.

papilla (pl. **papillae**): a minute, solid protuberance from the surface of a cells (especially of leaves and spores) of various forms, commonly domed or spinous, simple or branched. adj. **papillose**.

paraphyllium (pl. **paraphyllia**): a small, green, filiform, lanceolate or leaf-like scale borne superficially on the stems between branches of many pleurocarpous mosses, e.g.*Thuidium*; see also **pseudoparaphyllia**.

paraphyses (sing. **paraphysis**): sterile hairs composed of uniseriate cells, coloured or hyaline, associated with antheridia and sometimes archegonia.

parenchyma: tissue of undifferentiated cells, usually isodiametric and thin-walled, usually not overlapping; adj. **parenchymatous**.

paroicous: with antheridia and archegonia in the same gametoecium but not mixed, the antheridia immediately below the perichaetium in the axils of leaves.

patent: used for leaves spreading at an angle of about 45°.

patulous: used for leaves spreading widely (45–90°).

pedicel: a short stalk.

pellucid: clear, transparent or translucent.

pendant: drooping or hanging down, e.g. the capsules of *Bryum* ; or stems that hang, e.g. *Papillaria*. (= **pendulous**).

percurrent:*of a costa*, extending up to but ceasing at the apex of a leaf.

perfect: a complete peristome; applied to diplolepidous peristomes with an endostome having both segments and cilia.

perianth: inflated envelope surround the fertilized archegonium.

perichaetial leaf: a modified leaf surrounding the archegonia.

perichaetium: the female gametoecium, consisting of the sex organs and the perichaetial leaves surrounding them.

periclinal: oriented parallel (rather than perpendicular) to the surface. cf. **anticlinal**.

perigonial leaf: a modified leaf associated with and surrounding the antheridia.

perigonium: special bracts round the male flower.

Perigynium: the involucre of the female inflorescence in bryophytes.

peristome: a circular structure generally divided into 4, 8, 16 or 32 teeth arranged in single or double (rarely multiple) rows around the mouth of the capsule and visible after dehiscence of the operculum.

peristomial formula: an equation indicating the peristomial number from the outer peristomial layer (OPL) to the inner peristomial layer (IPL), and indicating relative degree of wall thickening and any lateral displacement of the IPL and prostomial development.

peristomial number: the number of cell columns in the outer, primary and inner peristomial layers per 45° arc (one-eigth peristome).

peristomial cylinder: the three innermost layers of amphithecial tissue in an arthrodontous moss capsule which produce the peristome. The inner peristomial layer is proximally continuous with the outer spore sac; the middle and outer layers represent the primary and outer peristomial layers, respectively.

phaneropore: a superficial stoma in a capsule wall having the guard cells on the same level as the exothecial cells. adj. **phaneroporous**.

phyllodioicous: with dwarf male plants growing on the leaves or tomentum of much larger female plants.

piliferous: with a long hairpoint.

Pinnate: feather like.

pinnate: with spreading branches on either side of a stem, rather like a feather.

pitted:*of a cell wall*, having small depressions or pores.

placenta: the interface between the gametophyte and sporophyte, usually containing numerous transfer cells. adj. **placental**.

plane: flat, not curved or wavy, as in leaf margins.

pleurocarpous: having sporophytes produced laterally on short, usually specialised branches rather than from the apex of the main stem; mosses with stems usually prostrate, creeping and freely branched, growing in mats rather than tufts. cf. **acrocarpous**.

plica: a lengthwise fold or pleat. adj. **plicate**.

plumose: closely and regularly pinnate and feathery in appearance.

polymorphic: having more than one form, variable.

polyploid: a plant or tissue with more than 2 complete sets of chromosomes.

polysety: having more than one sporophyte produced from a single gametoecium, each from a separate archegonium with its own calyptra, e.g. *Dicranoloma dicarpum*. adj.**polysetose**.

pore: a pit or opening in a cell wall. adj. **porose**.

portical lobe: belonging to the lower surface of thallus, stem or leaf.

primordial utricle: the collapsed contents of a cell that have separated from the cell wall.

process: the main divisions of a diplolepidous peristome (also called **segments**).

procumbent: prostrate, spreading.

prolate: longer than wide. cf. **oblate**.

propagule: a reduced bud, branch or leaf functioning in vegetative reproduction.

prora: a mammillose projection formed by protusion of the end of a prosenchymatous cell. adj. **prorate** ; dim. **prorulate**.

prosenchyma: a tissue consisting of narrow, elongate cells with overlapping ends. adj. **prosenchymatous**.

prostome: a rudimentary structure outside and usually adhering to the main peristome teeth; e.g. in Pterobryaceae.

prostrate: lying flat on ground; creeping.

protandrous: maturation of the antheridia prior to the archegonia.

protogynous: maturation of the archegonia prior to the antheridia.

protonema (pl. **protonemata**): a filamentous, globose or thallose structure resulting from spore germination and including all stages up to production of one or more gametophores. The protonema varies in the amount of chlorophyll present and the degree of obliqueness of its end walls, and in its branching.

protuberant: projecting.

proximal: the end or part nearest to the base or place of origin. cf. **distal**.

pseudautoicous: having dwarf male plants epiphytic on the female.

pseudoelaters: sterile cells mixed with spores in the capsule of *Anthoceros*.

pseudoparaphyllium (pl. **pseudoparaphyllia**): structures resembling paraphyllia but restricted to the bases of branches and branch buds in some pleurocarpous mosses.

pseudopodium: an elongation of the stem of the gametophore, e.g. below the sporophyte in *Sphagnum* and *Andreaea*, to give a false seta; also an extension of the stem tip bearing clusters of gemmae, e.g. in *Trachyloma*.

pseudopore: a pore-like structure with a thin membrane that is revealed by staining; e.g. in the hyalocysts of some Calymperaceae; in *Sphagnum* leaves consisting of fibril rings without an interior perforation.

pulvinate: cushion-like.

punctate: minutely dotted.

pyriform: pear-shaped, e.g. the capsules of *Bryum*.

Q

quadrate: *usually of cells*, appearing square or approximately so in two dimensions.

R

rachis: the axis of a pinnate or umbellate frond.

radial: spreading from a common axis or centre.

radiculose: covered with rhizoids.

ramification: branching.

ramose: richly branched.

receptacle: a structure with or without a stalk containing archegonium or antheridium.

recurved: curved down (abaxially) and inward, the opposite of **incurved**; in leaves referring to margins, apices or marginal teeth; in the peristome, teeth curved outward and ±downward.

reflexed: bent down (abaxially) and inward, the opposite of **inflexed**; generally referring to leaf margins or leaves of a stem.

reniform: kidney-shaped.

repand: with slightly uneven margin.

resorbtion: the digestion or erosion of cell walls in the leaves of some species of *Sphagnum*.

resorbtion furrow: a groove along the leaf margins of some species of *Sphagnum* caused by erosion of the outer cell walls.

reticulate: forming a network.

retort cells: cortical cells in some species of *Sphagnum*, with a downwardly projecting neck ending in a pore.

retuse: a slight indentation or notch in a broad, rounded apex.

revolute: of leaf margins, rolled downward (abaxially) and backward.

rhizautoicous: monoicous, with the male antheridiumon a short branch attached to the female plant by rhizoids and so appearing to be separate.

rhizoid: a hair-like structure that anchors a moss to the substratum; multicellular with oblique cross walls, often pigmented, and sometimes clothing the stem.

rhizoid furrow: a furrow in the stalk of the receptacle in the Marchantiaceae for covering the rhizoids from the receptacle.

rhizome: a slender horizontal, subterranean stem giving rise to erect secondary stems; e.g. in *Dawsonia* and *Rhodobryum*.

rhombic: diamond-shaped.

rhomboidal: longer and narrower than rhombic, oblong–hexagonal.

rostellate:*of an operculum*, with a short beak.

rostrate:*of an operculum*, with an apical beak that is narrowed to a slender tip or point.

rosulate: resembling a rosette, with leaves enlarged and crowded at the tips of stems.

rugose: with irregular, roughly transverse wrinkles or undulations; e.g. the leaves of *Neckera*.

rugulose: minutely or somewhat wrinkled transversely.

S

saccate: like a bag.

sacculate: like a small sack.

saxicolous: growing on rock.

scabrous: rough.

scale: a thin flat semitransparent plate of cells.

scarious: very thin and stiff like a scale.

scleroderm: a tissue of thick-walled cells in the central cylinder of stems and branches of *Sphagnum*.

secund: bent or turned to one side.

segment:*of a peristome*, a single, tooth-like component of the endostome.

seriate: in rows (uni-, bi-, tri- or multiseriate); applied either to adjacent rows of leaf cells, or to ranks of leaves on a stem. cf. **stratose**.

serrate: regularly toothed like a saw; leaves with marginal teeth pointing forward.

serrulate: minutely serrate.

sessile: without a stalk, e.g. of sporophytes with greatly reduced setae.

seta (pl. **setae**): the elongated portion of the sporophyte between the capsule and the foot.

setaceous: bristle-like.

setulose: bristle like.

sheathing: surrounding or clasping a stem, seta or capsule.

shoulder: the distal part of the leaf base where it is abruptly narrowed to the upper lamina or limb.

sigmoid: s-shaped.

sinuate: wavy.

sinuose: having a wavy wall or margin.

sinus: a gap between two lobes of a leaf.

slime-papillae: papillae with slime extracted from the swollen extremity.

spathulate: having the shape of a spatula, narrow below and gradually broadening above.

spermatozoid: a male gamete; bearing two flagella.

spiculose: sharply and minutely toothed or papillose.

spinose: having sharply pointed teeth.

spinulose: with minutely sharply pointed teeth.

splash-cup: a cup-shaped androecium in which the dispersal of antherozoids is aided by the action of falling raindrops.

spore: a minute, usually spherical, haploid cell produced in the capsule as a result of meiosis; its germination gives rise to the protonema.

spore sac: a spore-containing cavity in a moss capsule.

sporocyte: a diploid cell that undergoes meiosis in the capsule to produce 4 haploid spores; sometimes called a **spore mother cell**.

sporophyte: the spore-bearing generation; initiated by the fertilization of an ovum; consists of foot, seta and capsule; attached to and partially dependent on the gametophyte.

sporopollenin: a substance in moss spore walls similar to that found in pollen grains.

spreading: of leaves inserted at 46–90° to the stem; said to be widely spreading when close to 90°.

squarrose: *of leaves*, spreading at right angles to the stem.

squarrose-recurved: spreading at right angles, with the tips curved downwards.

stance: the manner in which the leaves are held in relation to the stem.

stegocarpous: a capsule with a differentiated, dehiscent operculum.

Stellate: star shaped.

stereid: a slender, elongate cell with very thick walls present in groups (**stereid bands**) in the costa and stem of many mosses.

stipe: the erect, unbranched basal part of a stem in a dendroid or frondose moss.

stolon: a slender, elongate branch with leaves that are often smaller and have a different shape to those of the main stem. adj. **stoloniferous**.

stoma (pl. **stomata**): a pore involved in gas exchange, surround by two guard cells; in mosses restricted to the neck of the capsule.

stratose: in layers; denoting the thickness of leaves, i.e. uni-,bi- or multistratose.

stria (pl. **striae**): a fine line or ridge. adj. **striate**.

striolate: very finely ridged.

struma: a cushion-like swelling at one side of the base of a capsule. adj. **strumose**.

Stylus: a small awl like lobule.

subula: a long, slender, needle-like point; adj. **subulate**.

substratum: the surface on which a moss grows, e.g. soil, bark or rock.

Succubous: the oblique insertion of the distichous leaves of liverworts so that the upper overlaps the lower on the dorsal side of the stem.

sulcate: with longitudinal folds or ridges, e.g. capsules of *Ulota*.

superficial: *of stomata*, having the guard cells in the same plane as the adjacent exothecial cells.

sympodial: having a main stem of determinate growth, and further growth by innovations or lateral branches.

synoicous: having antheridia and archegonia mixed in the same gametoecium.

systylious: *of a capsule*, the operculum remains attached to the tip of the columella after the capsule has opened.

T

terete: smoothly cylindrical, round in cross-section.

terricolous:

tessellate: checkered in little squares, applied particularly to the peristomes of some of the Tortulaceae.

tetrad: a group of four; e.g. the 4 spores derived from a single sporocyte by meiosis.

tetrahedral: a four-faced cell or spore.

thallus: a plant body not differentiated into stem anfd leaves, flat and broad like a frond.

theca (pl. **thecae**): the spore-bearing part of a moss-capsule.

tomentum: a felt-like or woolly covering composed of abundant rhizoids on some stems, rarely on leaves. adj. **tomentose**.

trabecula (pl. **trabeculae**): projecting cross-bars formed from the horizontal walls on either face of arthrodontous exostome teeth; also strands of cells bridging spaces within some capsules. adj. **trabeculate**.

transfer cells: specialised cells at the interface of the gametophyte and sporophyte which transfer nutrients from the former to the latter.

trigone: triangular intracellular wall thickenings found in the corners of three adjacent cells.

trilete spore: having a three-pronged scar on the wall (e.g. in *Sphagnum*), the scar being its area of contact with each of the three other spores in the tetrad.

trigonous: having 3 obtuse angle.

triquetrous: having three acute angles.

triradiate ridge: a thickening on the proximal face of a spore caused by it being pressed against the three other spores of a tetrad.

tristichous: arranged in 3 rows.

truncate: cut off abruptly or squarely at the apex.

tuber: a gemma borne on rhizoids, usually underground.

tubercles: peg like projections on the inner walls of the rhizoids in Marchantiaceae.

tuberculate: with small worts.

tuft: a growth form with stems erect but radiating at the edges and forming small cushions.

tumid: swollen or inflated.

Turbinate: top shaped.

turf: a growth form with stems erect, parallel and close together and forming rather extensive patches.

U

umbellate: a frondose moss having all of its branches spreading from the apex.

umbonate : convex with an abrupt, rounded central point.

uncinate: hooked; with the tip bent to form a hook.

underleaves: a third row of leaves on the ventral ride of the stem.

undulate: wavy.

uniseriate: arranged in one row.

unistratose: cells disposed in a single layer.

urceolate: urn-shaped; used with reference to capsules that are constricted below a wide mouth, then abruptly narrowed to the seta.

urn: the spore-bearing part of the capsule.

utricle: a bladder-like structure.

V

vaginant: one of two clasping leaf laminae in *Fissidens* spp.; the adaxial part of the leaf that sheathes the stem and encloses the base of the leaf above it.

vaginate: sheathing.

vaginula (pl. **vaginulae**): the sheath enveloping the base of the seta, derived from the basal part of the venter of the archegonium and surrounding stem tissue and remaining after the separation of the calyptra.

valve: one of the divisions of the capsule wall after dehiscence.

venter: the swollen basal part of an archegonium, containing the ovum.

vermicular: worm-like; long narrow and curving.

verrucose: covered with worm like protruberances.

verruculose: covered with wart-like protuberances.

verticillate: whorled.

W

weft: a loosely interwoven growth, often somewhat ascending.

whorled: arranged in a ring or circle.

widespreading: *of leaves*, spreading from the stem at a wide angle (less than 90°).

X

xerophyte: a plant that is adapted for survival in arid places. adj. **xerophytic**.

Z

zygote: the product of the fusion of male and female gametes; the fertilized ovum before it undergoes mitosis or meiosis.

References

Bold, H.C. 1957. Morphology of Plants, Harper & Row, New York.

Bold, H.C., Alexopoulos, C. and Delevoryas, T. 1987. *Morphology of plants and Fungi.* New York : Haper and Row, pp. 912.

Cavers, F. 1910-1911. Inter-relationships of the Bryophytes, I-X. *New Phytol.* 9 & 10.

Crandall-Stotler, B. 1980. Morphogenetic designs and a theory of bryophyte origin and divergence. *Bio Science,* 30 : 580-585.

Crandall-Stotler, B., Stotler, R.E. and Long, D.G. 2009. Morphology and classification of the Marchantiophyta, In: *Bryophyte Biology* (Eds.) B. Goffinet and A.J. Shaw, Cambridge University Press, pp. 565.

Crum, H.A. 2001. *Structural diversity of bryophytes.* University of Michigan Herbarium, pp. 379.

Drinnan, A.N. and Chambers, T.C. 1986. Flora of the Lower Cretaceous Koonwarra fossil bed (Korumburra group), South Gippsland, Victoria. *Association of Australasian Paleontologists, Memoir,* 3 : 1-77.

Duckett, J.G. and Renzaglia, K.S. 1988. Ultrastructure and development of plastids in bryophytes. *Advances in Bryology,* 3 : 33-93.

Duckett, J.H., Fletcher, P., Francis, R., Matcham, H.W., Read, J.T., Russell, A.J. and Pressel, S. 2004. *In vitro* cultivation of bryophytes; practicalities, progress, problems and promise. *Journal of Bryology,* 26: 3-20.

Dyer, A.F. 1979. The culture of fern gametophytes for experimental investigation. In: *The experimental biology of fens,* Ed. A.F. Dyer. Academic Press, London, pp. 253-305.

Eichler, A.W. 1883. *Syllabus der Vorlesungen uber spezielle and medicinisch-pharmaceutische Botanik,* 3rd Edn., 68. pp. Berlin.

Engler, A. 1892. *Syllabus der Vorlesungen uber spezielle and medizpharm. Bot.,* W. Engelmann, Leipzig.

Gangulee, H.C. 1966-1972. *Mosses of Eastern India and adjacent regions.* Fasc. I. 1966, Fasc. II. 1971, Fasc. III. 1972. Calcutta.

Gangulee, H.C. 1974-1977. *Mosses of Eastern India and adjacent region,* Vol. II (Fasc. 4-6), Books and Allied Limited, Kolkata, India, 831-1546 pp.

Gangulee, H.C. 1985. *Handbook of Indian Mosses.* Amerind Publishing Co. Pvt. Ltd. New Delhi.

Goffinet, B. and Shaw, J. 2009 (Ed.). *Bryophyte Biology*. Cambridge University Press, pp. 565.

Goffinet, B., Buck, W.R. and Shaw, A.J. 2009. Morphology, Anatomy and classification of the Bryophyta, In: *Bryophyte Biology* (Ed.) B. Goffinet and A.J. Shaw, Cambridge University Press, pp. 565.

Gradstein, S.R., Churchill, S.P. and Salazar Allen, N. 2001. A guide to the bryophytes of Tropical America. *Memoirs of the New York Botanical Garden*, 86 : 1-577.

Haeckel, E. 1866. *Erster Band: Allgemeine Anatomie der Organismen. Generelle Morphologie der Organisemen*. Verlag Van George Reimer, Berlin, 1-574 pp.

Hallingbäck, T. and Hdgetts, N. Status survey and conservation action plan for bryophytes – Mosses, Liverworts and Hornworts. Status survey and conservation action plan for bryophytes. IUCN/SSC bryophyte specialist group. IUCN, Gland, Switzerland and Cambidge, UK, pp. 106.

Howe, M.A. 1899. The Hepaticae and Anthocerotes of California. *Mem. Torrey bot. Club*. 7 : 1-208.

Jian-Cheng, Z., Shi-Liang, H., Min, L., Sulayman, M., Jie, H., Yuan-Ming, Z. and Xiao, L. 2004. A study on the characteristics of spore germination and protonemal development in *Lindbergia brachyptera. Arctoa*, 13: 223-228.

Kenrick, P. and Crane, P.R. 1997. *The origin and Early Diversification of Land Plants: A cladistics study*. Washington D.C.: Smithsonian Institution Press.

Krishnan, R. and Murugan, K. 2014. Axenic culture of bryophytes: A case study of liverwort *Marchantia linearis* Lehm. & Lindenb. *Indian Journal of Biotechnology*, 13: 131-135.

La farge-England, C. 1996. Growth form, branching pattern and perichaetial position in mosses: Cladocarpy and pleurocarpy redefined. *Bryologist*, 99: 170-186.

Meusel, H. 1935. Wuchsformen und wuchstypen der Europaischen Lanbmoose. *Nova ACTA Leopolding (N. F.) 3* (12): 124 – 277.

Mishler, B.D. and Churchill, S.P. 1984. A cladistics approach to the phylogeny of the "bryophytes". *Brittonia*, 36 : 406-424.

Mishler, B.D., Lewis, L.A., Buchhein, M.A. 1994. Phylogenetic relationships of the "green algae" and "bryophytes". *Annals of the Missouri Botanical Garden*, 81 : 451-483.

Nehira, K. 1988. Germination and protonema. In: *Methods in Bryology*, (Ed.) J.M. Glime, Nichinan, Hattori Botanical laboratory, pp. 113-117.

Olesen, P. and Mogensen, G.S. 1978. Ultrastructure, Histochemistry and notes on germination stages of spores in selected mosses. *The Bryologist*, 81: 493-516.

Papp, B., Odor, P. and Szurdoki, E. 2005. Methodological overview and a case study of the Hungarian Bryophyte monitoring program. *Bol. Soc. Esp. Briol.* 26-27.

Proskaur, J. 1957. Studies on Anthocerotales V. *Phytomorphology,*7 : 113-135.

Qiu, Y.L., Li, L., Wang, B. 2006. The deepest divergenses in land plants inferred from phylogenomic evidence. *Proceedings of the National Academy of Sciences*, U.S.A., 103: 15511-15516.

Renzaglia, K.S., Schuette, S., Duff, R.J. 2007. Bryophyte phylogeny : advancing the molecular and morphological frontiers. *Bryologist*, 110 : 179-213.

Renzaglia, K.S., Villareal, J.C. and Duff, R.J. 2009. New insights into morphology, anatomy and systematics of hornworts, In: *Bryophyte Biology* (Ed.) B. Goffinet and A.J. Shaw, Cambridge University Press, pp. 565.

Rothmaler, W. 1951. Die Abteilungen und Klassen der Pflanzen, *Feddes Report*. 54 : 256-266.

Rowntree, J. K. and Ramsay, M. M. 2009. How bryophytes came out of the cold: successful cryopreservation of threatened species. *Biodivers Conserv*, 18: 1413 – 1420.

Sabovljević, M., Vujičić, M., Pantović, J. and Sabovljević, A. 2014. Bryophyte conservation biology: *In vitro* approach to the *ex situ* conservation of bryophytes from Europe, *Plant Biosystems – An international journal dealing with all aspects of plant biology*, 148:4, 857-868, DOI : 10. 1080/11263504. 2014. 949328.

Schimper, W.P. 1879. Palaeophytologie. In: K.A. Zittel. *Handbuch der Palaeontologie*. 2, Lief. 1: 1-152.

Schofield, W.B. 1985. *Introduction to Bryology*, Caldwell, NJ : Blackburn Press.

Schuster, R.M. 1953. Boreal Hepaticae, A manual of the liverworts of Minnesota and adjacent regions. *The American Midl. Nat.*, 49 : 257-684.

Schuster, R.M. 1958. Annotated key to the orders, families and genera of Hepaticae of America north of Mexico. *The Bryologist*, 61 : 1-66.

Singh, D., Dey, Monalisa and Singh, D.K. 2010. A synoptic flora of Liverworts and hornworts of Manipur. *Nelumbo*, 52:9-52.

Stotler, R. and B. Crandall-Stotler, 1977. A check list of the liverworts and hornworts of North America. *Bryologist,* 80 : 405-428.

Takhtajan, A.L. 1953. Phylogenetic principles of the system of higher plants. *Bot. Rev.* 19 : 1-45.

Thomas, B.A. 1972. A probable moss from the lower carboniferous of the forest of Dean, Gloucestershire. *Annals of Botany,* 36 : 155-161.

Tuba, Z., Slack, N.G. and Stark, L.R. (Ed.) 2011. *Bryophyte Ecology and Climate Change,* Cambridge University Press, pp. 503.

Van Aller Hernick, E.L. and Bartowski, K.E. 2008. Earths oldest liverworts – *Metzeriothallus sharonae* sp. Nov. from the middle Devonian (Givetian) of eastern New York, U.S.A. *Review of Palaeobotany and palenalogy,* 148 : 154-162.

Van der Poorten, A., Papp, B. and Gradstein, R.2010. *Sampling of Bryophytes.* In: Eyman, J., DeGreef, J., Hauser, C., Monje, J.C., Samyn, Y. and Vanden Spiegel, D. (Eds.). *Manual on field recording techniques and protocols for all taxa biodiversity inventories and monitoring,* Vol. 8, pp. 331-345.

Villarreal, J.C., Cargill, D.C., Hagborg, A., Söderström, L. and Renzaglia, K.S. 2010. A synthesis of hornwort diversity: Patterns, causes and future work. *Phytotaxa,* 9:150-166.

Walton, J. 1925. Carboniferous bryophyte II. Hepaticae and Musci. *Annals of Botany,* 42 : 707-716.

Warming, E. 1896. *Lehrbuch der ökologischen Pflanzengeographaie.* Bornträger, Berlin.

Watson, E.V. 1981. *British mosses and liverworts.* Cambridge University Press, New York, pp. 519.